T0143131

Manufacturing
Systems

Manufacturing Systems

An introduction to the technologies

Second edition

D. J. WILLIAMS

Head of Department
Professor of Manufacturing Processes
Department of Manufacturing Engineering
Loughborough University of Technology

KLUWER ACADEMIC PUBLISHERS
DORDRECHT / BOSTON / LONDON

Published by Kluwer Academic Publishers,
P.O. Box 17, 3300 AA Dordrecht, The Netherlands.

Sold and distributed in North, Central and South America
by Kluwer Academic Publishers,
101 Philip Drive, Norwell, MA 02061, U.S.A.

In all other countries, sold and distributed
by Kluwer Law Academic Publishers,
P.O. Box 322, 3300 AH Dordrecht, The Netherlands.

First edition 1988
Reprinted 1991
Second Edition 1994, Reprinted 2001

ISBN 0 412 60580 5

A catalogue record for this book is available from the British Library
Library of Congress Catalog Card Number: 94-71201

Printed on acid-free paper

Printed in the Netherlands

To Katy and Elizabeth

Contents

Preface

I have updated this second edition to take account of more recent understanding of the complementary roles of people and computers within manufacturing. The manufacturing technologies described in the first edition of the book (which was revised a little on its reprinting) remain as relevant, to both the Western and Japanese manufacturing economies as they were in 1987. We do now however have a better understanding of how these technological tools can be applied as part of an overall manufacturing strategy. Recent CIM applications have shown us that while computers are, axiomatically, essential tools for competitive manufacturing, we should only use them after the complexity of the manufacturing task has been reduced and where the enterprise and the computer systems within it are capable of accommodating the changes forced upon them by external pressures.

A number of additions have been made to the book, the further reading has been completely revised and the book now closes with a set of example questions to support the text. I have also added a number of bullets at the start of each chapter to indicate to the reader the particular focus of that chapter.

This revision of the book has been encouraged by my publisher, Mark Hammond of Chapman & Hall. I would like to thank him and the publishers of the first edition for the care and guidance that they have given me in preparing both editions of the book.

The book, when first published, built upon much work by the Open University and the Open University Press; it should be acknowledged that many of the figures, particularly in the robot chapter, arise from this source. The front cover and Figure 8.17 are reproduced with the permission of Megamation. Also since the publication of the first edition of the book a number of companies that supplied figures have changed their names including Asea to ABB and KTM to FMT.

I hope that this small, and hopefully accessible, book continues to be of value to the manufacturing community.

David J. Williams,
Loughborough, 1993.

Preface to the first edition

It is essential for the traditionally industrialized countries to innovate in manufacturing to survive in the increasingly competitive world marketplace. This challenge coupled with the increasing application of computers has led to significant changes in the techniques applied in manufacturing. This book seeks to introduce those technologies that are being applied in discrete parts manufacturing.

In the technical press there have been many phrases and acronyms coined to describe these technologies including numerical control (NC), machining centres, computer aided manufacture (CAM), computer integrated manufacture (CIM), simulation, robotics, flexible manufacturing systems (FMS), automatic assembly, factory automation, Kanban, just in time (JIT), manufacturing automation protocol (MAP), advanced manufacturing technology (AMT), etc. The book is intended to introduce senior undergraduates, postgraduate students and practising engineers to the principles of these individual technologies and their integration into complete, automated, programmable manufacturing facilities and systems. It is hoped that this will allow the reader to have a critical perspective of the marketplace and potential solutions to his own current or future problems. It is also intended to indicate how the complete manufacturing facility can be viewed as a system.

The book does not address the related areas of computer aided design (CAD), scheduling, production control and current speculative research at any significant level. It is impossible to do justice, in this short book, to such large subject areas which, without doubt, demand books in their own right.

A book such as this is still necessarily wide-ranging and occasionally superficial. Consequently the text is supported by a bibliography of relevant books and papers so that individual topics, including many not covered in the text, can be pursued in depth.

The industrial (and first time) reader should pass rapidly over the analytical elements of the book – these, though principally aimed at an academic audience, are intended to encourage a more analytical approach to a largely anecdotal subject.

David J. Williams,
Cambridge, 1987.

Acknowledgements

The author would like to thank the following organizations specifically for allowing the reproduction of illustrations and analysis:

Aims for Industry, Bedford, UK; Asea Manufacturing Systems, Luton, UK; Asea Robotics, Milton Keynes, UK; ASME, New York, USA; John Brown Automation, Coventry, UK; British Robot Association, Kempston, UK; Build Group, CIT, UK; Cahners Publishing, London, UK; Cincinnati Milacron, Birmingham, UK; Cross International, Liverpool, UK; Dynapert Precima, Colchester, UK; DEA, Turin, Italy; Fanuc UK, Ruislip, UK; 600 Fanuc, Colchester, UK; GEC Electrical Projects, UK; Gould, Basingstoke, UK; George Fisher, Schaffhausen, Switzerland; GMF Robotics, Troy, USA; IBM Robotics, Crawley, UK; Istel, Redditch, UK; Jungheinrich, Manchester, UK; KTM, Brighton, UK; Makino Milling Machines, Atsugi, Japan; McDonnell Douglas, St Louis, USA; McGraw Hill, Peterborough, USA; Meta Machines, Abingdon, UK; MIT Press, Cambridge, USA; National Bureau of Standards, Gaithersburg, USA; NC Graphics, Cambridge, UK; Prentice-Hall International, Englewood Cliffs, USA; Renishaw, Wooton under Edge, UK; Rolls Royce, Derby, UK; Rover Group, Cowley, UK; Siemens, Congleton, UK; TI Machine Tools, Coventry, UK; Unimation, Telford, UK; Visolux Ltd, Cambridge, UK.

Such a list can never be complete. There are many references in the text, figures and bibliography to installations and other work – details of which have been widely published. The designers and authors of these must also be thanked and are acknowledged where possible.

The author would like to thank all his former students and his colleagues in the Manufacturing Group at Cambridge for the many contributions that they have made to the book. Particular thanks are due to Paul Rogers for his detailed and extensive comments on the manuscript, and to Ann Barber and to Jo Tree for their assistance in editing the typescript.

1

Introduction – manufacturing systems approaches

After reading this chapter the reader should understand:

- the structure of the book;
- the business drives on manufacturing;
- manufacturing systems approaches;
- the place of single machines within manufacturing systems.

1.1 Introduction

The increasing application of computers has led to significant changes in the techniques applied in manufacturing. This book seeks to introduce those technologies that are being applied in discrete parts manufacturing and aims to make the reader familiar with systems approaches to the design and operation of conventional and programmable manufacturing systems. More traditional approaches to the organization of manufacturing operations are briefly discussed to indicate the context of the new technologies.

An emphasis on the control, programming and monitoring of both single machines and machine systems is essential to the understanding and implementation of the newer systems approaches, together with an understanding of the place of people within systems. This book therefore concentrates on these aspects as well as the more traditional production engineering emphasis on facility and factory layout. The book addresses particularly the use of programmable manufacturing facilities.

The book covers traditional systems configurations, the control and mechanics of single machines such as machining centres, robots and machine sensor systems and the software for single machines. The integration of these single devices into various examples of manufacturing cells is examined, together with the enabling technologies for this integration. We then take these cells and further integrate them into manufacturing systems. The book presents examples of such systems together with the computer architectures and software systems used to control them – briefly addressing the fundamentals of simulation for system design where appropriate.

The text introduces each of the individual concepts in sufficient depth to allow the reader to understand how they fit together to create the complete manufacturing facility. Because such a treatment is necessarily superficial, it is supported by a Further Reading section of relevant books and papers (at the end of the book) so that individual topics can be pursued in greater depth.

1.2 The structure of the book

This chapter outlines current systems approaches to manufacturing problems, indicating that the factory can essentially be viewed as a system synthesized from sub-systems. Each of the sub-systems in a factory interact in a complex manner, as do the elements of each sub-system. The chapter also sets out the goals that the factory has to fulfill and shows why these have increasingly led to the introduction of programmable automation.

Chapter 2 begins to examine the manufacturing facility sub-system in detail. Here, conventional approaches to factory layout and the strategies of dedicated automation are outlined, together with simplifying philosophies such as Kanban and JIT.

1.3 Single machines

The complete manufacturing facility is a very complex system – and before applying the complete technology one should have experience of the individual technologies. Elements of the system are also commercially and technically significant in their own right. The body of the book therefore begins by examining the building blocks of programmable facilities rather than complete systems. Some of the fundamental building blocks are described in Chapters 3–6.

We begin by examining the CNC machining centre, the servo-controlled device that has set the pace for programmable systems. It also provides a good example of a programmable processing machine, where traditional machine tool technologies have had to be exploited to bring the machine to where it is today. The CNC machining centre is of special significance as it is often applied in isolation to give significant improvements in operating efficiency.

Chapter 4 is dedicated to the programmable manipulator, the robot, which handles parts between individual processing machines and carries out some of our manufacturing processes – and begins to create our automated system. It is also increasingly applied in the assembly of products. The types of robots encountered and the fundamentals of their most significant task – manipulation of parts in three-dimensional space – is indicated.

In programmable manufacturing facilities it is necessary to use a great deal of sensing. Chapter 5 examines simple sensing and the basics of perhaps the most presently applicable high level sensor, vision. Vision is important because the technology can be applied in isolation, and also allows immediate feedback of product quality changes to the automated manufacturing system.

After concentrating on the 'mechanics' of each device, trying to indicate why and how they work as they do, Chapter 6 then discusses the programming of devices – especially on how they are programmed to attain a position in space. Robots and NC machine tools are considered together in this chapter to indicate their similarities. Throughout the book lines of program ('code') are shown with a simple English language 'translation'. The code and the chapters on programming are not included to enable people to program any of the devices – they are intended to make the reader familiar with (and unafraid of) the many languages that they will encounter in programmable automation.

1.4 Cells, assembly and systems

Chapter 7 begins the examination of the integration of discrete devices into manufacturing cells, regarded by many as the building blocks of large systems and applicable, individually, as islands of automation. The chapter concentrates on the mechanical arrangements of the cell and includes a number of real examples of the application of the technologies. The chapter closes by examining CAD and logical models of cells.

Chapter 8 turns to assembly – particularly important in manufacturing – and which does not readily yield to programmable automation. This chapter outlines the mechanics of the insertion process, design for assembly principles, and briefly examines the design of dedicated assembly machines and programmable assembly cells.

The next three chapters examine the complete programmable manufacturing facility (automated factory) and address the arrangement of the system, the computer control of the system and system software respectively.

Chapter 9, which discusses the manufacturing facility as a hierarchy and the concepts of flexibility and reconfigurability, examines briefly the technologies of GT, DNC and flexible automation systems – with examples from industry worldwide, indicating their areas of application. This chapter also takes a practical view of the necessity for people in automated systems and a brief look at discrete event simulation.

The final chapters examine the computer control of such facilities as an integrating, enabling technology. Chapter 10 introduces hierarchical control and computer architectures for manufacturing control, while Chapter 11 deals with system control software. The book closes in Chapter 12 with a number of example problems.

1.5 The manufacturing process

The book does not address the manufacturing process itself – the actual shape or property-changing activity that adds value to a component or assembly. There are a number of reasons for this: it is particularly difficult to generalize about processes and they all take considerable development before they can be successfully automated. Manufacturing processes also require the application of considerable mechanical engineering ingenuity to develop and solve particular tooling and fixturing problems. Such developments have to be driven by the particular product being manufactured. Automation cannot proceed successfully until such development has taken place. The content of this book is intended to be applicable over a wide range of products, and therefore concentrates on the more generally applicable techniques for the control and programming of the manufacturing facility and the machines within it.

1.6 Business drives for manufacturing

There are a number of different drives that have led to the current developments in manufacturing technology and approaches to manufacturing problems. These can essentially be divided into two groups: economic pressures and more general commercial issues.

Economic pressures have led to a particular drive to reduce work-in-progress and inventory and the size of manufacturing facilities. Inventory is the stock held in warehouses, both of incoming raw materials and sub-assemblies, and work-in-progress is the stock held between manufacturing processes on the shop floor. These stocks represent large amounts of idle capital that could be better employed, and has led to an emphasis on the construction of facilities with fast door-to-door time and just-in-time (JIT), minimum inventory, minimum paperwork production philosophies. In the past there was a similar emphasis on 'floor to floor time' – the length of time that a component was being processed on a machine. This was optimized without particular attention being paid to its optimal and fast progress through the complete manufacturing facility – the 'door-to-door time'.

Commercial market pressures have had a number of effects too – customers now demand increased product variety. Compare this with the (perhaps apocryphal) saying of Henry Ford, 'You can have any colour as long as it is black!', which permitted a production and factory control operation of considerably less complexity. Customers also expect shorter delivery and lead times. These have resulted in the reduction of batch sizes, which are traditionally associated with economic production (see the discussion in Chapter 2), placing increased emphasis on the construction of more flexible and programmable facilities that are easier to change from the manufacture of one product to another.

In industries where there is a very large variety of parts and a requirement for very small batches (such as the aerospace and prototyping industries) 'batch of one' philosophies for production facilities have been proposed. The theoretical batch of one facility is capable of processing a workpiece presented as a single unit amongst other different workpieces, with the necessary hardware and software changes being achieved at minimum cost. It is therefore capable of making any batch size at its optimum performance and can produce any workpiece on demand with a short lead time. As a result the costs associated with set-up need to be at a minimum, and this pushes the technology towards totally programmable machines which have no mechanical changeover costs.

Another significant market pressure has been consumer demand for increased quality. This, combined with the cost of the inspection process in most conventional facilities has led to the development of a 'right-first-time' philosophy. This implies that the older strategies of 'inspecting quality into the product' and process development by reducing the quantities of scrap produced are replaced by actually making the product correctly in the first place by understanding the capability of the process! One of the major sources of variability in product quality is the human operator, who tires and gets bored. Most programmable automation is introduced to give better, consistent quality and significant increases in quality by operating in constant conditions. The 'right-first-time' approach encourages a continuous effort to improve standards by always reducing the number of rejects that are acceptable. The electronics industry is a particularly striking example of this, as it strives to reduce the failures in soldered joints to single figure proportions per million joints.

Throughout the discussion there is a demand for improved control of a situation with increasing complexity – not only control of the particular process that changes the shape of our product but of the whole manufacturing facility. This has led to the development of manufacturing systems approaches that consider this whole manufacturing facility.

There has been a significant drive in recent years towards the reduction of the number of people involved in manufacturing operations because of the high cost of labour. This was particularly true when manufacturing organizations had large numbers of operators recruited when the atmosphere of international competition was more favourable. Over the last few years the number of operators in these facilities has been trimmed or the companies operating them have closed. Much of the automation that we will discuss cannot therefore be justified on manpower savings alone. The high total cost of the introduction of the new technologies is such that the cost difference between using one operator or two to run a large facility becomes less significant when compared with the total investment. It is also rarely appreciated that a system operator also carries out many 'invisible' tasks, such as inspection or component orientation which are far too costly to automate. We should also recognize that people within manufacturing systems also carry out many integration tasks, such as peer-to-peer information transfer, that are difficult and costly to implement using an automated system. Such an automated integration system is also likely to be expensive to re-engineer when the product or process changes.

1.7 Manufacturing systems approaches

Manufacturing systems approaches seek to optimize the initial design to commercial product time, the design lead time and factory door-to-door time, the manufacturing lead time, by considering the whole factory as a system and simplifying and optimizing the performance of this complete system. Recent approaches have emphasized that the whole of the manufacturing enterprise, including for example the marketing and logistics activities, must be considered in this optimization. It is also likely to be necessary to consider interactions with other companies, such as suppliers, to determine how the extended enterprise serves its final customer.

One of the key considerations in this optimization of the system performance is that the measure of that performance is correct. In a capitalist, industrialized society such as ours, where the manufacturing process represents adding value and wealth creation (as indicated in Fig. 1.1) global performance measurements are immediate profit and long term

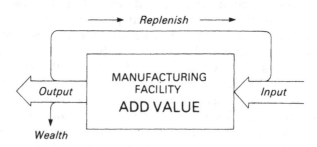

Fig. 1.1 Wealth creating by manufacturing
Source: Lupton (1986)

competitiveness. These performance measurements can be determined by examining the performance of world best practice competitors – using the achievements of others as benchmarks to drive your own results in increased manufacturing effectiveness.

Let us briefly consider the definition of a system, and look at some of its properties. A system is an assembly of components or sub-systems organized so that each sub-system is connected to each other component. These components are affected by being in the system and the behaviour of the system is changed if a sub-system leaves it.

It is also widely assumed that a system is synergistic; in other words, the system has more properties than the sum of the properties of its parts. A human analogy is often presented to justify this: a human being is considered to be more than the sum of its organs.

Business pressures have led to the development of two sorts of manufacturing systems approaches. One approach is essentially a 'top down' approach and is more amenable to systems analysis and perhaps management tasks, while the 'bottom up' functional integration approach is perhaps better suited to systems synthesis as it is a more familiar engineering approach.

Both of these are the hard manufacturing systems approaches. There are many fields that use systems approaches. Some of these take a hard approach, considering that any system can be modelled in terms of its inputs, outputs and its internal flows and conversions. Figure 1.2 shows a technical system, and the flow of material, energy and information. The flows of material and information are the most significant in the consideration of manufacturing systems. Information should also be considered to embody signal and control information, and such information may be transferred by people as well as by electronic means.

A 'soft' systems approach, on the other hand, uses a technique which constructs a particularly abstract 'ideal' system model and then compares the actual system with this to highlight areas that need change.

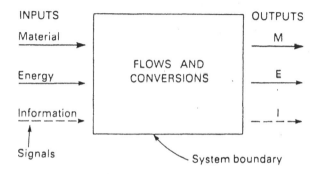

Fig. 1.2 A technical system

1.8 Top down approaches

Figure 1.3 shows a model for a manufacturing system proposed by Parnaby (1979) which introduces less tangible business inputs and outputs to the purely technical system that was outlined above. Some of these, such as social pressures and company reputation, are not directly under the control

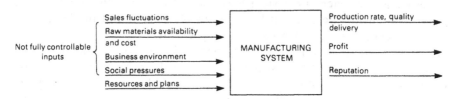

Fig. 1.3 An input–output analysis for a manufacturing system
Source: Parnaby (1979)

of the company. This model describes the context in which the manufacturing organization operates. If we move within the company (moving from the top, down) this model can be extended to describe an actual production sub-system (Fig. 1.4). You will see that this includes all the familiar functions of the production engineering activity within a factory – the

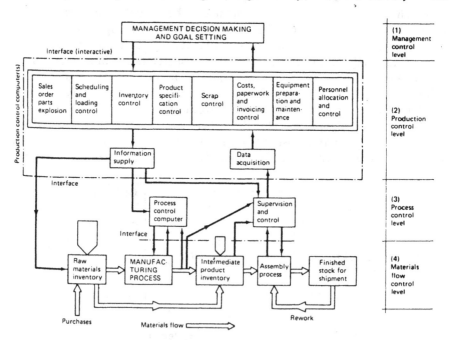

Fig. 1.4 Levels of control and information flows in a production sub-system
Source: Parnaby (1979)

manufacturing hardware, the material flows (purchases and rework) and the information flows (manual and computer control data). This very tangible system model is similar to that produced by a bottom up approach.

The technique of beginning by modelling the whole company as a collection of sub-systems of inputs and outputs, and then modelling each of these sub-systems as a further collection of sub-systems is a widely recognized systems analysis tool. The IDEF0 modelling system is such a method. This is based on the incremental block (Fig. 1.5), which has inputs,

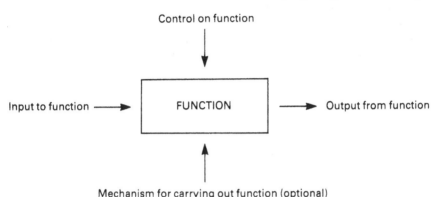

Fig. 1.5 The IDEF0 model block ·

outputs and control functions – and this last function is usually shown as coming from above. The mechanism which indicates how the function of the block is carried out is rarely shown. Starting from a single initial block,

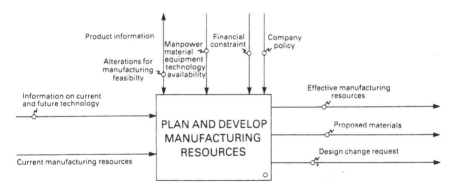

Purpose: To investigate the interfaces between major function areas as in AO diagram.
To define detailed information flow between low level function.
To identify who does each low level function.
To identify duplication of effort.

Viewpoint: Production Engineering Manager

Fig. 1.6 An IDEF0 model
Source: Cracknell

each block within a system is expanded into three further blocks. The expansion by threes helps to restrict the amount of expansion, and addition of unnecessary detail that is done at any one time. An IDEF0 model of manufacturing resources within a company is shown in Figs 1.6 and 1.7, indicating three way expansion and the links between the blocks. The IDEF0 committee work is based on a structured design method for developing software, SADT (Structured Analysis and Design Technique) for large computer programs. Such programs are large and unwieldy systems that need to be modelled and divided into sub-systems, to allow a number of programmers to work on different elements of the problem at the same time and give a coherent solution. Such diagrams are sometimes known as HIPPO – hierarchical input output – diagrams.

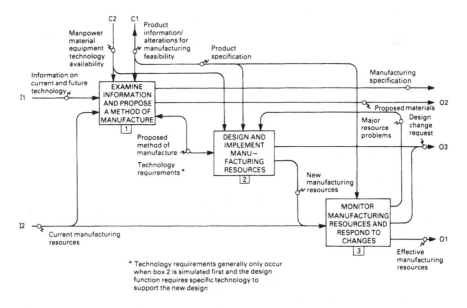

Fig. 1.7 The first expansion of the IDEF0 model
Source: Cracknell

If a company is modelled like this, the structured approach highlights inconsistencies in company practices and organization systems. Manufacturing facilities are much easier to synthesize with a bottom up, functional integration approach because the actual processes used in the manufacture are fixed.

1.9 Bottom up approaches

In the bottom up approach the system is synthesized from its elements, where the exact description of the inputs and outputs and the characteristics of the flows and conversions are known. This is obviously similar to the concepts used in the design of automatic control systems where the systems are represented by inputs, outputs and transfer functions. Figure 1.8 shows a model of an 'ideal' computer integrated manufacturing (CIM) system synthesized from presently available practicable elements. The reader will recognize that the diagram does not include support of the marketing activity. This view of CIM also tends to over-emphasize the role of the computer within the organization – we must understand that the computer is only a tool that we use in our primary task, that of integrating the organization to serve its customer.

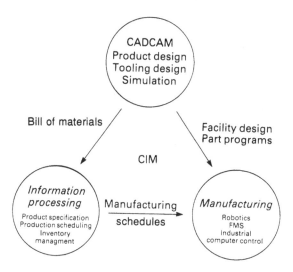

Fig. 1.8 A view of CIM
Source: Saville

There are many definitions of CIM, ranging from 'the computer assistance to all engineering and business functions from order entry to product shipment' to those that actually encompass the integration of the facility, and the implications that this has for the passing of data between all sub-systems. By any definition such a model is not at present actually achievable from its elements. This is because there are limitations on the technologies and their interfaces. The model also takes no account of the requirements for feedback in the system – nor of the influence of people on the system. It is interesting to explore how these considerations affect the model.

In synthesized systems, it is usually considered that the synergy within such systems comes from information sharing between sub-systems.

1.10 Automation and people

Let us take a bottom up, functional integration approach in describing the elements and synthesis of programmable manufacturing facilities. It has only recently become possible to take such a view in a wide range of manufacturing facilities as a result of the enabling technology of inexpensive distributed computing using microprocessors. Within manufacturing it is usual to set a 'top down' goal for the organization and to take a bottom up view of how we might implement a system to allow the achievement of that goal.

In this book we are essentially addressing the automation of manufacturing facilities. Automation is the technology concerned with the application of mechanical, electronic and computer based systems in the operation and control of production – generally resulting in the replacement of manual tasks by machine.

Throughout the book we will take a very mechanistic view of the role of people in the manufacturing process. We must not, however, lose sight of the fact that the issue of people is emotive and that we must retain some compassion in the consideration of our function. We must also ensure that all the high technology devices that we wish to apply are configured so that they are transparent – easy to use – to anyone that is required to use them. And no matter how novel or interesting a piece of technology is, it serves no purpose in the practical world unless it is financially justified.

2 More conventional approaches to factory layout

After reading this chapter the reader should understand:

- traditional factory layouts;
- group technology;
- dedicated automation strategies;
- just in time and continuous improvement philosophies.

2.1 Introduction

In this chapter we will start by examining the more traditional forms of factory and some of the methods that are used to design them. We will then outline some of the techniques used in dedicated automation. The closing section of the chapter looks at the application of the Kanban technique, a simple and powerful technique for factory control to push facilities towards minimum inventory and just in time production.

2.2 Traditional factory layouts

There are essentially four types of layout for the conventional manufacturing facility:

- 'static' or 'fixed' position build,
- process-based layout

and product-based layouts, i.e.

- flow lines
- continuous production.

These are discussed more fully below.

2.3 Static build

Static or fixed position build is most often encountered in the assembly of very complex structures, such as aeroplanes or high quality machine tools. Both of these applications involve the production of a very small number of complex structures which are based around a large starting point, such as the base casting or fabrication of a machine tool or the fuselage of an aircraft.

The process often involves the fitting of components, where final finishing of the component parts may actually be carried out at the build site. The characteristic of the process is that there are a large number of components in the final assembly that are small (and often difficult) to put together. Many tradesmen work on the same product and these highly skilled people come to the product – the product does not go to the people as it does in most higher volume production.

2.4 Process-based layout

The most traditional and frequently encountered form of factory has a process-based layout where machines of a certain type are grouped together (Fig. 2.1). The rough machining operations are carried out in the area nearest to the incoming parts store and the finish machining is carried out in an area which is nearest to the finished parts store. If the ingoing and outgoing materials transport areas are combined, the flow of material follows a 'U' shape around the factory. Such facilities produce a wide variety of parts, in one-off or small batches, and use a large amount of skilled labour. Large transport and work-in-progress penalties are usually associated with older facilities of this sort which have grown over a number of years, indicated in

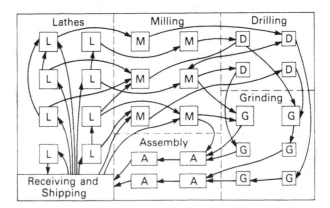

Fig. 2.1 A process layout
Source: Groover (1986)

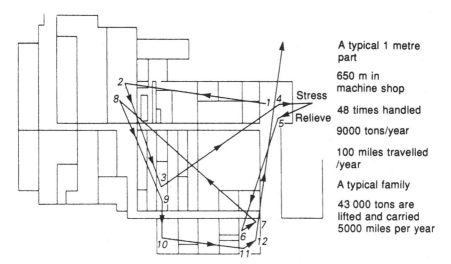

Fig. 2.2 The problem of organic growth in a process layout
Source: Lupton (1986)

Fig. 2.2. The variability of the skill and amount of attention paid to the manufacturing task by an operator leads to quality variation.

The layouts outlined rely on expeditors (or progress chasers) to push jobs through the factory. Such methods are notoriously inefficient as they generally result in the progressing of jobs, apparently with the most urgency, in an *ad hoc* manner.

2.5 Product-based layouts

Most other facilities have product-based layouts, making one single product, or very closely related products. Examples of this are the chemical plant, which makes a single set of products (such as detergents) by a continuous, uninterrupted, process and the automotive production line, for the assembly of cars. Such facilities allow very sophisticated factory control and automation systems to be installed because the constrained product range simplifies the problems of factory system design. The systems produced when automated give significant improvements in quality. This increase in quality is associated with the standardization and closer tolerances required for automation.

2.5.1 FLOW LINE BASED MANUFACTURING SYSTEMS

Flow line based factory layouts were pioneered by Henry Ford in the early part of this century by his development of the assembly line, where partially completed assemblies on a conveyor (often known as a 'track') pass operators who add their components to gradually complete the assembly. Cars had been previously built by static build, with a small number of highly skilled tradesmen. By dividing the assembly process between many operators (with each operator carrying out a small, less skilled element of the task as determined by a scientific management or Taylorist (after F. W. Taylor) approach) and by bringing the work to the operator Ford speeded up the manufacturing process – and thus became less dependent on the skill of his employees. (Scientific management sought to divide each task into sub-tasks, the performance of which could be performed in a prescribed way.)

In the flow line the product travels serially from process to process or machine to machine, as is indicated in Fig. 2.3. This method, originally applied for assembly, led in turn to the development of the purpose built transfer line (or transfer machine) for component machining, in which components are passed along a line of machines by a dedicated transfer mechanism or conveyor.

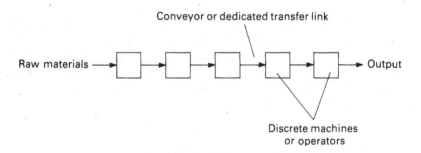

Fig. 2.3 The flow or transfer line

A further advantage of line based approaches for manual tasks, such as assembly, is that the line paces the operator at a constant, reasonably predictable speed by bringing parts continuously.

Recent years have seen much discussion and some rejection of the scientific management approaches – task sub-division and operator de-skilling – that are implicit in the above. The application of continuous improvement philosophies – where the operator continually improves the effectiveness of the way he performs his job – presupposes that the operator has a wider understanding and commitment than just with his immediate task. These sorts of purpose built or dedicated automation systems are sometimes known as 'Detroit automation' as they were largely pioneered and exploited in the United States' motor industry. The problem with such large, dedicated, often (in the early days) purely mechanical, systems is that they are very difficult to design to produce a variety of products.

When, for example, a model change has taken place it has been necessary to replace most of the installation.

We will examine the transfer line and dedicated automation philosophies a little later in the chapter.

2.5.2 CONTINUOUS PROCESS LAYOUT

This layout is perhaps best exemplified by the continuous chemical plant (such as an oil refinery) where the whole of the plant has been specifically designed and optimized to carry out what is essentially a single task. Because of this the plant has a very complex and sophisticated control system which allows close control of quality. Most newer types of factory layout attempt to give the control associated with such layouts while retaining some product flexibility. It is outside the terms of reference of this book to talk extensively about continuous process plants as they are not applied in the discrete parts manufacturing activity.

The selection of the type of layout that is most suitable is carried out using a break even analysis such as that represented by Fig. 2.4, which plots anticipated production cost and sales revenue against output quantity per unit of time. The cross over points indicate at which output volumes the layout choices become applicable.

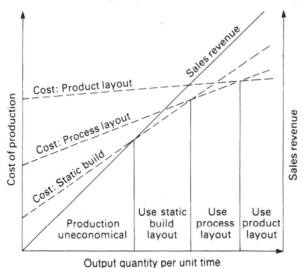

Fig. 2.4 The break even analysis
Source: Wild (1980)

2.6 Techniques of factory layout

The factory layout task – laying out machines in sections, sections in departments, departments in factory – is essentially one of iterative design in which there are many variables that must be considered. Table 2.1 gives an indication of the items that must be laid out.

Table 2.1 What has to be laid out in the factory?

Machines and position of old plant	Services:
Support structures – gantries and mezzanines	Swarf/scrap
Foundations	Coolant
Handling – conveyors and cranage	Tooling
People – offices and benches	Hydraulics
Gangways – AGVs and forktrucks	Air
Catwalks	Power
Guarding	Control
Maintenance access	Steam
Heating and lighting	Lubricants
Storage and WIP buffers	
Parts input and output	

In many commercial organizations, the layout activity for a new product will almost always start with a fairly well defined process plan (the method of manufacture of the part). However, the machine supplier and its exact specification may not have been decided, and consequently the first task is then to decide on the type of machine. The factors affecting this are:

- process,
- machine capacity,
- machine size,
- machine speed,
- degree of mechanization.

It should be recalled at this stage that the manufacturing activity should be able to feedback suggestions to changes in the product design that will allow the manufacturing activity to be easier.

The process plan to be considered in the facility design stage for optimization can be represented using work measurement symbols as an aid (Fig. 2.5). These represent actual operations (circles), transport (arrows), delays (Ds), storage (triangles) and quality control activities (squares). These are work measurement symbols pioneered by the 'scientific managers' for work study and allow the representation of tasks and the times allotted to those tasks. By identifying the actual activities that contribute to the adding of value to the product – the operation – and the more redundant operations – transport and delays – they allow the easier optimization of the process plan and route of the part through a manufacturing facility. Such practices have much less significance today than they did in the early years of their invention.

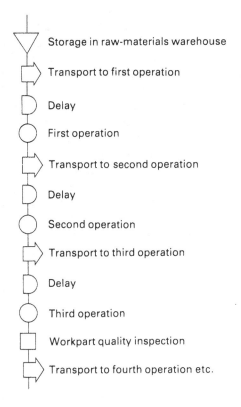

Storage in raw-materials warehouse

Transport to first operation

Delay

First operation

Transport to second operation

Delay

Second operation

Transport to third operation

Delay

Third operation

Workpart quality inspection

Transport to fourth operation etc.

Fig. 2.5 Work measurement representations

Firstly, it is necessary to define the work holding and tooling (jigs, fixtures and tools) at this stage so that the transport and storage facilities for these can be identified and included in the design. (The chief function of a jig is to guide a cutting tool and the chief function of a fixture is to hold a part.) Next, identify the number of machines required and the buffer sizes between them by considering the following factors:

- cycle times,
- down times,
- setting times,
- transport times,
- control rules.

The preferred solution is derived through iterations of the above process using analytical tools like assembly line balancing or modelling, using discrete event simulation which is discussed in Chapter 9. See Wild (1980) for a description of the operational research (OR) tools that have been developed to assist the process.

Buffers are necessary between machines to reduce downtime in 'linked lines' – if there are no buffers the whole line must shut down when one machine stops. Downtime is non-productive time on a machine caused by either the breakdown of the machine itself or breakdown of another machine that causes the machine to shut down. Large buffers, however, give high work-in-progress penalties.

The task is to now place and move the machines on the shop floor in two dimensions. This is essentially an iterative and intuitive process. The handling between machines, for the components themselves and for services such as swarf and coolant, is then placed. One of the factors requiring critical optimization at this stage is the transport distances between machines. This can be done using a 'string' diagram (Fig. 2.6) to minimize the complexity and length of these paths. String diagrams are often annotated with work measurement symbols to show inconsistencies in a layout. Machines at this stage are usually represented by their envelope in plan, the travel of any slideways on the machine and their interconnection details.

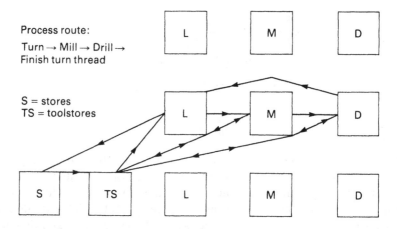

Fig. 2.6 A string diagram

The final task is to lay out the preferred solution, if it is particularly complex, in three dimensions, either by conventional drafting, CAD geometric simulation (see Chapter 7) or by using physical models.

2.7 Group technology

We will discuss in this section and sections 2.8 and 2.9 the use of group technology, dedicated automation, just-in-time (JIT) and Kanban to further increase the efficiencies of conventional manufacturing operations.

In a group technology (GT) layout, different sorts of conventional machines are grouped together into 'GT cells' (Fig. 2.7). Recall from biology that a cell is essentially the smallest autonomous unit. Each cell is capable of making a small variety of similar parts, the parts being handled between machines manually and the machines themselves being controlled by an operator. Unautomated GT cells are often arranged as flow lines to avoid congestion.

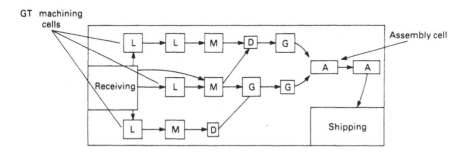

Fig. 2.7 A GT layout.

The parts made within the cell are selected to be similar by either classification and coding or by visual inspection. It is important to recognize that this selection, although based on shape, collects together components that are made by similar process routes. In the coding method parts are classified by a complex code, the digits of which represent, for example, whether the part is cylindrical or prismatic, whether it has changes in diameter or a slot, etc. There are many types of codes, and an example of part of the code developed by Optiz is shown in Fig. 2.8. In the visual inspection method, the parts are put into groups by looking at them 'in the flesh', or at drawings or photographs. This is only possible in facilities manufacturing a small number of components. Many industrialists consider the second approach as efficient as the first. A third approach is to say 'our product is pumps, therefore we need GT facilities that make pumps' – thus the pump becomes the group. An example of two groups of parts as manufactured in the Scamp facility at the 600 Group is shown in Fig. 2.9. These are both cylindrical groups, one of shafts and one of discs.

Group technology cells can reduce work-in-progress and generally increase the operating efficiency of small batch manufacture by reducing handling and transport costs. The design discipline implied by the grouping activity also reduces the proliferation of very similar but different product designs which fulfil essentially the same function and assists production planning.

	Part class	External shape, external shape elements	Internal shape, internal shape elements	Plane surface machining	Auxiliary holes and gear teeth
0	Rotational parts: $\frac{L}{D} \leq 0.5$	No machining	No machining: Without bore, without through hole	No plane surface machining	No gear teeth: No auxiliary bore
1	Rotational parts: $0.5 < \frac{L}{D} < 3$	Smooth, no shape elements	No machining: With through hole	External plane surface and/or surface curved in one direction	No gear teeth: Axial holes without indexing
2		Stepped to one end or smooth: No shape elements	Stepped to one end or smooth: No shape elements	External plane surfaces related to one another with a pitch	No gear teeth: Axial holes with indexing
3		Stepped to one end or smooth: Screwthread	Stepped to one end or smooth: Screwthread	External groove and/or slot	Axial and/or radial holes and/or in other directions
4		Stepped to one end or smooth: Functional groove and/or functional taper (and screw thread)	Stepped to one end or smooth: Functional taper (radial groove and screw thread)	External spline (polygon)	Axial and/or radial holes with indexing and/or in other directions
5		Stepped to both ends (multiple increases): No shape elements	Stepped to both ends (multiple increases): No shape elements	External spline, slot and/or groove	Spur gear teeth without auxiliary holes
6		Stepped to both ends (multiple increases): Screwthread	Stepped to both ends (multiple increases): Screwthread	Internal plane surface and/or groove	Spur gear teeth with auxiliary holes
7		Stepped to both ends (multiple increases): Functional groove and/or functional taper (and screw thread)	Stepped to both ends (multiple increases): Functional taper (radial groove and screw thread)	Internal spline (polygon)	Bevel gear teeth
8		Operating thread	Operating thread	Internal spline, external groove and /or slot	Other gear teeth
9		Others	Others	Others	Others

Fig. 2.8 GT coding
Source: Optiz

2.8 Dedicated automation*

There are a number of strategies used by the designer of automation to increase the effectiveness of the manufacturing operation. These are applicable across all fields of automation, but here we will consider their application largely in dedicated automation. Dedicated automation is taken to mean that the automation is engineered and constrained to only accommodate either a single part or a very small variety of parts and that it has no programmable elements.

2.8.1 SPECIALIZATION OF OPERATIONS

One method of increasing manufacturing productivity is to use a special purpose machine designed to carry out one single operation to the greatest possible efficiency. This is mainly a dedicated automation strategy. The chucking lathe is an example of this – it has no tail stock and is fed along the axis of the chuck rather than perpendicular to this axis, as is a conventional general purpose centre lathe.

* This section is adapted from Groover, M. P. (1980) *Automation Production Systems and Computer Aided Manufacture*, Prentice-Hall, Inc., Englewood Cliffs, NJ, pp. 35–37. Reprinted with permission.

Fig. 2.9 Groups of shaft parts and disc parts

2.8.2 COMBINATION OF OPERATIONS

The conventional production process is essentially a large number of distinct sequential steps on different machines. The objective of combining operations is to reduce the number of machines required by doing more than one operation on a given machine. This dramatically reduces set-up and work-in-progress.

2.8.3 SIMULTANEOUS OPERATIONS

This extends the philosophy by carrying out the combined operations simultaneously. An example of this is the use of multiple drilling heads. It is obvious that there are limitations to this philosophy because it usually requires purpose built tooling and is most usually used in dedicated automation. However, special purpose machines called head changers are often used in flexible manufacturing systems (FMSs) to reduce the cycle time required to carry out a large number of individual drilling operations. They use a multihead tool dedicated to each component to carry out simultaneous drilling operations.

2.8.4 INTEGRATION OF OPERATIONS

In this strategy a series of workstations are linked into a single integrated mechanism by automatic workhandling devices to form a transfer machine. Such machines are often used for metal cutting in automotive engine component manufacture. Typical linking mechanisms are walking beam conveyors and rotary tables, shown in Figs. 2.10 and 2.11. With several stations, more than one part can be processed by the system at any one time, reducing work-in-progress and total throughput time. There is a disadvantage to this system – if one station stops the whole line stops unless buffer stations or reject tracks have been designed into the machine.

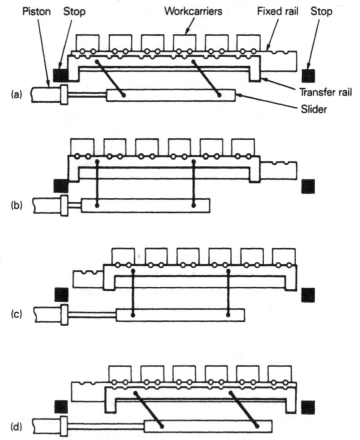

Fig. 2.10 A transfer line mechanism
Source: Boothroyd (1981)

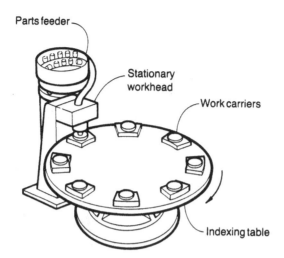

Parts feeder

Stationary workhead

Work carriers

Indexing table

Fig. 2.11 A rotary indexing table
Source: Boothroyd (1981)

2.8.5 REDUCE SET-UP TIME

Consider the reduction in time for each set-up – this can be accomplished by scheduling similar workpieces through a production machine and using common fixtures for similar but different parts (this is very similar to the GT approach). Figure 2.12 shows an example of this (projected by Rolls Royce, UK) to permit the rapid fixturing of a range of turbine discs. The production machines have the same fixtures mounted on them and the particular components are tailored to these using 'slave' rings and centres which fit into the fixture held on the machine.

2.8.6 IMPROVED MATERIALS HANDLING

Attacking the reduction of non-productive time by using mechanized materials handling methods reduces work-in-progress and throughput times. Such benefits are achieved, for example, by minimizing distances of travel and ensuring smooth flow, i.e. the use of conveyors and accumulators. Figure 2.13 shows a sketch of an accumulator chain. When a machine at the exit end of the accumulator chain shuts down, the pulleys in the centre of the accumulator will fall to effectively increase the length of chain between the machines. A component is usually held at a regular number of chain pitches and this extension therefore allows a buffer of components to form a linked line. When the machine at the exit starts to run again the chain will shorten.

The application of automated guided vehicles (AGVs) and robots is an example of improving materials handling by using programmable devices.

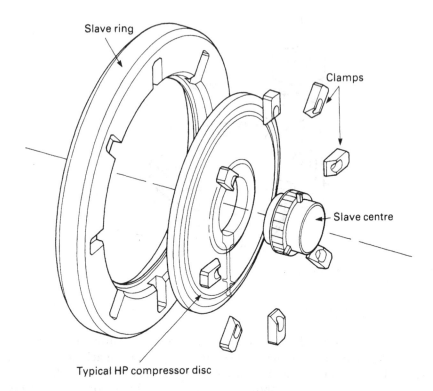

Typical HP compressor disc

Fig. 2.12 The use of 'slave' components to tailor parts to use common fixtures. *Source*: Rolls Royce

2.8.7 PROCESS CONTROL AND OPTIMIZATION

Improved process control will give a number of improvements, especially in quality. Examples of the application of this sort of strategy are early servocontrollers on NC machines which automated the 'take a cut – measure' cycle of the fully manual machinist. NC will be discussed at length in the next chapter.

2.8.8 COMPUTERIZED MANUFACTURING DATABASE AND CONTROL

These are automation strategies that are implemented with the computer as programmable automation of the manufacturing, planning and design process. They are therefore considered together with computer-based approaches to process control and materials handling.

Groover considers that manufacturing lead time (MLT) can be represented as

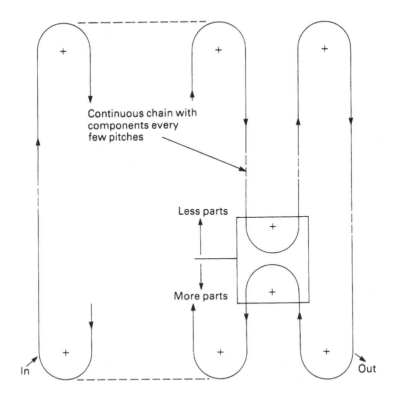

Fig. 2.13 An accumulator.

$$\text{MLT} = N_m \, (T_{su} + Q \, T_o + T_{no}), \qquad (2.1)$$

where N_m is the number of machines through which the product must pass, T_{su} is the set-up time, Q is the batch quantity, T_o is the operation time per machine and T_{no} is the non-operation time for each process – T_o being given by

$$T_o = T_m + T_h + T_{th}, \qquad (2.2)$$

where T_m is the actual process time, T_h is the workpiece handling time and T_{th} is the tool handling time per workpiece.

Groover examined the effect of the automation strategies on each of these elements of the lead time and the results are shown in Table 2.2.

2.9 Just-in-time and Kanban

When we adopt group technological approaches we are constraining and simplifying the complexity of the manufacturing task. Where we install dedicated automation we necessarily have to simplify and constrain the manufacturing task to make it possible. This simplicity has many advantages. We are now going to describe Kanban, a tool that can be applied when the manufacturing task has been simplified, and force it to improve its effectiveness.

Table 2.2 The effect of the automation strategies on manufacturing lead time
Source: Groover (1986)

Strategy	Manufacturing function	Objective to reduce
Specialization of operations	1	T_m, T_o
Combined operations	1,2	T_h, T_{th}, N_m
Simultaneous operations	1,2	T_h, T_m, T_{th}, N_m
Integration of operations	1,2	T_h, N_m
Reduce set-up time	1,3	T_{su}
Improve materials handling	2	T_{no}
Process control and optimization	1,3	T_m
Computerized database	4	N_m, T_{no}
Computerized control	3,4	T_{no}

Manufacturing functions to which these apply

1. Materials processing and assembly
2. Materials handling
3. Control – process and plant level
4. Manufacturing database development

The Toyota Motor Company has developed a powerful production control technique and manufacturing discipline called just-in-time. It is applicable in companies where the range of products produced is not excessive, and has been successfully applied in the motor industry worldwide. It has two main effects – it dramatically reduces inventory and work-in-progress and gives 'pull through' production control. Pull through implies that the customer drives the production operation directly – the production operation only making what the customer wants rather than making production as predicted by a forecast of the customer's requirements. This reduces overproduction and surplus capacity in the departments that precede final assembly and dispatch. Just-in-time seeks to set up units within the company that have customer–supplier relationships, the final customer in the plant being the assembly operation.

Just-in-time is intended to be implemented without a computer. This results in small data processing costs, and the simple nature of the Kanban makes it easy to collect production management data.

In JIT a 'Kanban' is an order card attached to a container of parts. Such cards take two forms; a 'conveyance Kanban' which is carried when going from one process to the preceeding process, and a 'production

Kanban' which is used to order the production of the parts used by the subsequent process.

When the contents of a container of parts begin to be used the conveyance Kanban is removed from the container and taken to the stock point of the preceding process and attached to the container held there. This container holds the production Kanban, and this is removed and used to trigger the production of parts to replace those withdrawn. These are produced as soon as possible. The processes in the line are therefore linked to the preceding processes or external subcontractors. By attention to the variables governing the Kanban movement, the operation of the plant is driven towards that of an indexing conveyor. This gives a just-in-time (JIT) and minimum inventory strategy and the manufacturing lead time is much reduced. The line and flow of Kanban and material is shown in Fig. 2.14.

The performance of a Kanban system can be evaluated using the following expression. Let y be the number of pairs of Kanban cards (the number of Kanban containers), D the demand per unit time, T_p the processing time, C the container capacity (not more than 10% of the daily requirement), and α a policy variable (not over 10%). Then

$$y = \frac{D(T_w + T_p)(1 + \alpha)}{C} \tag{2.3}$$

These factors are not regarded by Toyota as fixed values, but rather as targets for improvements. The factor α is a function of the facilities' ability to control external influences (a measure of the strength of the customer–supplier relationship), and in Japan this is regarded as a powerful measure of the performance of the plant by top managers. α is the amount of extra

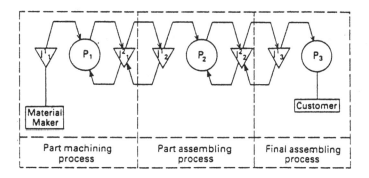

Fig. **2.14** The flow of parts and Kanban:
P$_i$ is operation or process i
1^1i   is part inventory for process i
1^2i   is finished goods inventory for process i
Source: Sugimori *et al.* (1977)

inventory in the process that has been included to buffer production against unforeseen events. D is determined from the consideration of smoothed demand and the number of Kanban is fixed, to an extent, by the processes involved and the design of the product – as D rises it is therefore necessary to reduce $(T_w + T_p)$ by reducing overtime and line stoppages by concentrating on tool change times and preventative maintenance. Work-in-progress is reduced by concentrating on C, y, and $(T_p + T_w)$.

Continuous improvement, Kaizen, techniques are used to drive the workforce to reduce inventory and breakdown times. In Kaizen, once a particular target is reached the target is then changed to require a more demanding performance. In this way the workforce continuously improves its performance while getting rewards for reaching intermediate targets. This process is usually associated with problem ownership and employee empowerment: the ability to identify problems, to take decisions and implement changes locally. This is in direct contrast to the traditional approach, promoted by scientific management, in Western companies where process design and improvement has been a 'white-collar' task detached from the day-to-day experience of running the manufacturing facility.

Kanban is a method of driving a plant to behave as a conveyor with no buffers, i.e. the just-in-time situation (in this case $C = 1$, T_w and $\alpha = 0$).

It is interesting to briefly contrast this with the materials requirements planning (MRP) and manufacturing resource planning (MRP II) approaches. Both of these systems are large software packages which are usually used to plan manufacturing predictively in situations of high complexity and over long time scales.

Philosophies like this do not attack the problem of reducing the overall complexity of the manufacturing control operation. Many consider that this complexity should be reduced by careful attention to detail before the implementation of any software system. This will allow simple control tools to be used, e.g. short term predictive MRP, combined with pull through Kanban philosophies, instead of large MRP II systems.

We now turn to the detailed examination of the programmable manufacturing facility in the rest of the book, beginning by examining a programmable process. We should, however, recall that all the programmable machines that we apply are only of value if the process is right, the conventional production engineering is right and if the facility we construct fits into the factory and company as a whole.

3 The machining centre – a servo-controlled machine tool

After reading this chapter the reader should understand:

- the importance of servo-control to modern manufacturing;
- concepts of accuracy and repeatability;
- how the machining centre has been configured for automated manufacture;
- the mechanical constraints on metal cutting machines.

3.1 Introduction

The application areas that can exploit programmable automation to its fullest extent are based on a controllable and 'programmable' process. Such areas at present include metal cutting with CNC machine tools, sheet metal nibbling with CNC machines, spot welding with industrial robots, electronic assembly using programmable placement machines and two-dimensional mechanical assembly with robots. In each of these cases the process has been developed so that it can be carried out by a computer-controlled device that can be programmed to carry out the particular task, but is flexible enough to produce a wide variety of different components or assemblies.

In this chapter we will examine the machining centre, a particular example of a programmable processing machine. The programmability of the metal cutting machine tool has led to many of the advances we are familiar with today, including the introduction of CADCAM systems and flexible manufacturing systems. The machining centre will be considered as an NC machine tool, and a tool for minimally manned manufacture. It will also be considered as a mechanical system where the process itself interacts with the machine tool to demonstrate the influence of the manufacturing process on the machine carrying out that process.

Machining centres have been specially constructed to be suitable for unmanned operation for periods of about one eight hour shift. Such machines are most frequently encountered in milling, drilling and boring operations, where they are called horizontal or vertical machining centres (depending on the direction of the axis of the cutter spindle) and for turning, where they are called turning centres. Turning centres often include facilities for small milling operations, and occasionally other more unusual processes such as grinding. A horizontal machining centre is shown in Fig. 3.1.

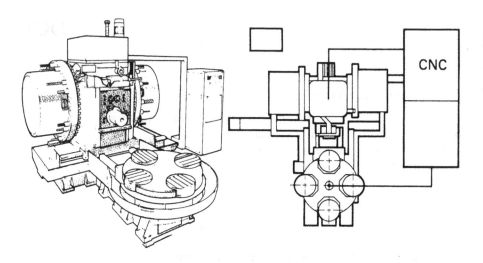

Fig. 3.1 A horizontal machining centre
Source: KTM

After briefly discussing servo-control, we will examine the horizontal machining centre in some detail, and look at two machine tool design problems which result in machine deflection and loss of accuracy. These are vibration leading to regenerative chatter and thermal distortion.

3.2 Feedback control

Programmable machines are driven by a control program to a position in space, and this position in space must be achieved consistently. Servo control is used to achieve this.

The axes of the two most commonly encountered programmable automation devices, the machine tool and the robot, have essentially the same feedback servo-control hardware. This feedback control technique, when used for machine tools, is usually known as numerical control (NC).

Such machines are moving objects in space under computer control, either as manipulators or as a tool cutter path. A simple machine tool consists essentially of a two-(x,y) or three-(x,y,z) Cartesian axis table being manipulated in front of a single point or multipoint cutter by a 'centre lathe' or 'milling machine'. A robot is a more complex device consisting of four or more axes (which are not necessarily Cartesian) and are linked and interact in a complex manner.

The axes of these machines at present are driven usually by DC servo motors and feedback is collected from the rotation of the servo motor, usually measured directly or indirectly using an encoder (robots and machine tools) or resolver (machine tools); its speed is measured by a tachogenerator.

Direct current motors allow a precise control of their speed over a wide operating range by variation of the voltage applied to the motor. There are hydraulic alternatives to the DC drive options which are used for large and powerful devices – these, however, have the usual disadvantages of hydraulics, such as oil leaks and the requirement to maintain an additional technology.

The feedback devices are used to measure the actual position and speed achieved so that they can be compared to the required position and speed inputs to the system. The most usually encountered feedback device is the incremental rotary encoder. This essentially consists of a transparent disc marked with an alternate clear and opaque pattern (Fig. 3.2). A light

Fig. 3.2 A rotary encoder

source is placed on one side of the disc and this is observed by a photocell which counts the number of light–dark transitions. Absolute encoders are sometimes used on robotic devices to avoid errors due to faulty data because of incorrect counts. These have multitrack encoder discs which define the position of the shaft absolutely using a binary word or other code (Fig. 3.3). The speed of devices can be collected by a tachogenerator, which is essentially a small permanent magnet DC generator whose output is proportional to its speed.

A schematic of a microprocessor-based implementation of such a system is shown in Fig. 3.4. Microprocessors which use digital rather than analogue data necessitate ADCs (analogue to digital converters) and DACs (digital to analogue converters) when using analogue sensors and actuators.

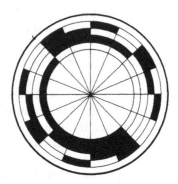

Fig. 3.3 The binary disc of an absolute encoder

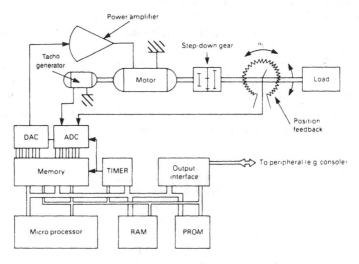

Fig. 3.4 A microprocessor implementation of servo control
Source: Coffret

There are three concepts that are used to measure the performance of programmable machines.

- *Resolution*. The smallest increment of distance that can be read and acted upon by an automatic control system. This is a function of the encoder increments.
- *Repeatability*. A measure of the positional error when returned to a trained or taught position under the same conditions (for example, the same direction of approach). This is of significance for devices that are programmed by teach (see Chapter 6) such as current industrial robots.
- *Accuracy*. A measure of the absolute positional capability – the difference between the actual position response and the target position desired or commanded by an automatic control system. This is most

important for NC machine tools and assembly robots because they can be programmed off-line (see Chapter 6).

Figure 3.5 shows the commonly understood concepts of accuracy and repeatability together with the more rigorous measures of system performance.

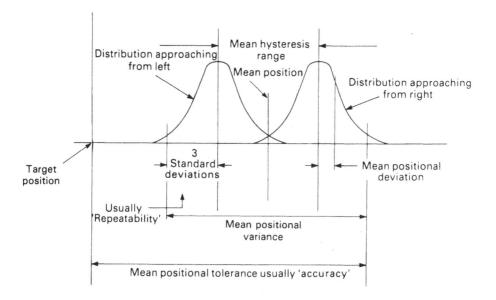

Fig. 3.5 Accuracy and repeatability

3.3 The horizontal machining centre

The horizontal machining centre is a sophisticated 3-(or occasionally more) axis milling, boring and drilling machine to make prismatic or box-like components (see Fig. 3.6). Horizontal implies that the spindle axis is horizontal. A lathe makes revolute or cylindrical components.

It has a number of features.

1. It can make components to high levels of accuracy. If closer tolerances can be achieved in the cutting operation it is likely that previously necessary finishing operations (such as boring or grinding) do not need to be carried out. This requires very high machine stiffness.
2. It is programmable – essentially minimizing the set-up time and tool change time previously required by an operator, and the servo control required for programmability ensures that the quality variations encountered in manually operated machines do not occur.
3. The machine is capable of being programmed off-line, using tapes or

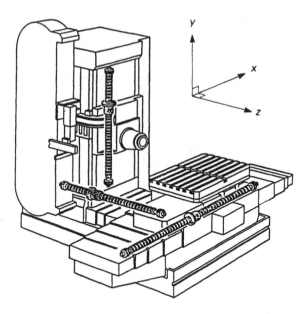

Fig. 3.6 The three major axes of a machining centre

direct numerical control (DNC) systems (these are discussed more fully in Chapters 6 and 9).

These requirements result in the use of CNC (computer numerical control) where the machine has its own resident control and programming computer. This CNC controller is usually able to collect management information.

4. The machine is required to run unmanned for extended periods. This requires the provision of:

  (a) a tool magazine, capable of carrying at least the tools for a single component (a typical tool magazine will contain 40 tools, although this may only cover the requirements for one single prismatic component);

  (b) a tool changer, to permit these tools to be inserted into the machine spindle (a tool changer is shown in Fig. 3.7);

  (c) a pallet changer (see Fig. 3.8) or pallet loop (see Fig. 3.9), to allow a variety of components that have been previously set up on pallets by an operator to be presented to the machine;

  (d) the CNC controller, to be able to store and manipulate a number of part programs (the program files defining the tools (cutters) used by the machine and the path they follow) and to be able to monitor the machine condition and shut the machine down when these conditions are unacceptable;

  (e) a swarf removal mechanism; metal cutting generates a lot of swarf,

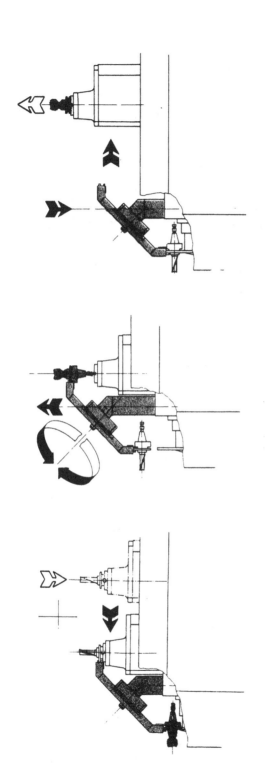

Fig. 3.7 A tool changer and tool magazine

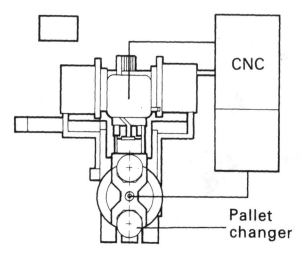

Fig. 3.8 A pallet changer
Source: KTM

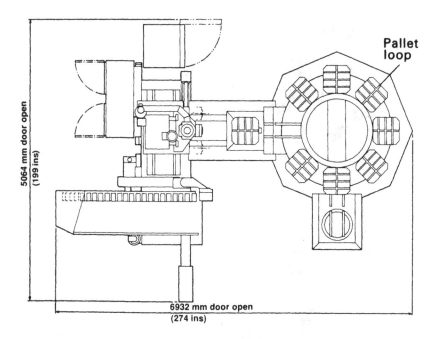

Fig. 3.9 A pallet loop
Source: Cincinnati Milacron

so this has to be removed. Swarf removal is not a trivial task as swarf strands tend to bind together. Metal cutting systems cannot be introduced without reliable swarf handling. This places emphasis on free machining materials and chip-breaking tooling. Swarf removal is often carried out by magnetic conveyors or Archimedian screws;

(f) a coolant system; swarf removal and the minimizing of machine temperature variation leads to the use of large volumes of coolant (cutting oil and water) which must be handled efficiently to and from the machine;

(g) a sensory capability; the machine also requires a variety of sensors to recognize parts presence, part position and tool breakage (see Chapter 5 for a discussion of the sensors that can be used in manufacturing).

These factors are equally true for turning centres, which have essentially the same requirements even if the embodiment of the design is different. A number of cells including a turning centre are included in Chapter 7.

A machine is usually utilized to maximize machine up time (provided that this is sensible within the context of the whole system operation). Machine up time has two elements: the cutting time and the tool change and set-up time. Many of the features above essentially seek to reduce the set-up and tool change times. By optimization of the overall manufacturing system performance, the waiting time (the time the machine is idle because it has no raw materials to process) can be reduced.

3.4 Constraints on the operation of machine tools

Engineers have traditionally worked to reduce the cutting of a component by increasing metal removal rates. Metal removal rates are limited by elastic and thermal distortion of the machine and cutting tool, leading to loss in accuracy and instability between the cutting process and the machine. The traditional machine tool disciplines have optimized the machine tool mechanical elements to their present sophistication. Without this sophistication and the associated high machine accuracy that gives a well-behaved machine and process the other newer programmable technologies would have had less impact. It is therefore of value to review some of these traditional technologies.

3.5 Elastic and thermal effects in machine tools

When high forces are applied by the cutting process to machine tools, the machine will deflect (as shown in Fig. 3.10) and lead to 'loss of form' on the finished component. This can be avoided by careful design of the machine to minimize the linear and rotational deflection.

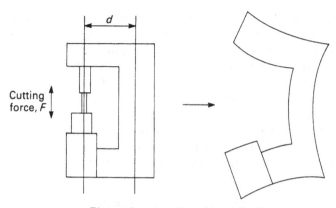

The cutting operation gives deflection

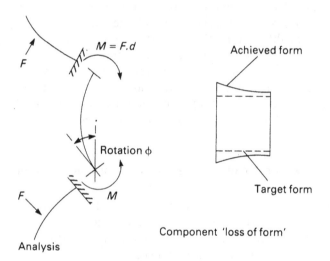

Fig. 3.10 The problem of elastic deflection

It is also important in maintaining the accuracy of machine tools to ensure that thermal distortion of the device is minimized – again giving rise to linear and rotational losses in accuracy. Such thermal effects arise from various heat sources, such as:

1. surrounding objects, e.g. heaters, walls, doors;
2. sun rays; and
3. the high or low temperature of coolants and lubricants input from separate sources such as hydraulic power packs.

In these cases heat transfer is mainly by radiation and convection (except 3). There are also more significant internal heat sources, the heat from which is transferred through the structure by conduction, such as:

4. drive motors (especially cutting spindle drives);
5. friction (in drives, lead screws and bearings); and
6. the cutting process itself, increased metal removal rates give more waste heat.

These effects must be minimized by design or can be software-compensated from measurements of the machine temperature. Variations in machine geometry are usually compensated for by software calibration of the NC system.

Design considerations which help to minimize these effects are:

1. external mounting of drives (see the y-axis drive in Fig. 3.1);
2. machine isolation from external sources, e.g. the application of jig borers in separate air conditioned areas;
3. careful dissipation of frictional heat from drives and bearings;
4. removal of process heat by coolant at a controlled temperature.

Thermal distortion effects are rotational and linear (see Fig. 3.11). The rotational effects can be particularly damaging, as a small rotational error can be magnified into a significant linear one by the lever arm through which it acts. Software compensation can at present only accommodate linear errors.

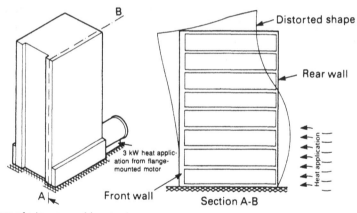

Fig. 3.11 Thermal distortion in a machine tool
Source: Weck (1984)

3.6 Regenerative chatter in machine tools

To understand regenerative chatter (the catastrophic deterioration of the cut surface due to vibration of the cutting tool), let us consider a machine tool as a vibrating system, where variations in the cutting force produce waviness in the cut surface. This waviness changes the dynamic cutting

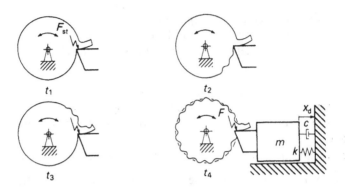

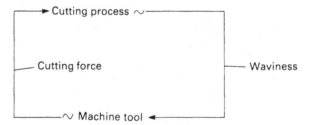

Fig. 3.12 Regenerative chatter in turning

force, which in turn excites the machine tool. A schematic of this effect is shown in Fig. 3.12 for the turning process. The factors which affect this dynamic cutting force are most important as they establish the stability of the cutting process itself and hence define its limits. The dynamic cutting force is mainly dependent on the changes in cross section of the chip being cut, i.e. on the variations of the active cutting tool edge length and the chip width.

The following conservative analysis indicates some of the factors affecting this behaviour, and allows an expression for the critical chip width (at which regenerative chatter will commence) to be determined.

Consider the system shown in Fig. 3.13 and the cutting process shown in Fig. 3.14. In the diagram, X is the mode direction of the machine (the direction in which it vibrates as a single degree of freedom system), P the cutting force direction, and Y the surface vibration of interest.

```
  ┌──────► Cutting process ~─────────┐
  │                                   │
  │                                   │
  │─ Cutting force        ───── Waviness
  │                                   │
  │                                   │
  └────────~ Machine tool ◄──────────┘
```

Fig. 3.13 The cutting system

From the diagram the chip thickness variation is

$$Y - Y_0 \text{ or } Y_0 - Y. \tag{3.1}$$

Now, resolving the cutting force and the chip thickness variation in the mode direction of the machine (as shown in Fig. 3.15) $\Delta P'$, the variation of the cutting force in the mode direction, becomes

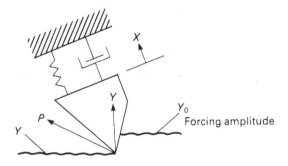

Fig. 3.14 The cutting process

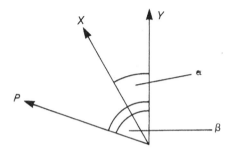

Fig. 3.15 The resolution of forces and directions

$$\Delta P' = \Delta P \cos (\beta - \alpha), \tag{3.2}$$

and

$$Y = X \cos \alpha. \tag{3.3}$$

Now let us examine the behaviour of the machine as a single degree of freedom system of mass m, stiffness k, and damping coefficient c, under a forcing vibration approximated to a sine wave

$$P = P \sin \omega t,$$

and the system behaves to give a displacement of

$$X = x \sin (\omega t - \phi),$$

where ϕ is the phase lag between the forcing and forced vibration. The velocities and accelerations are therefore

$$\dot{X} = x\omega \cos (\omega t - \phi)$$

and

$$\ddot{X} = - x\omega^2 \sin (\omega t - \phi),$$

and the velocity can be represented as

$$\dot{X} = x\omega\,[\,j\,\sin\,(\omega t - \phi)].$$

This last expression is obtained by considering the velocity vector being rotated through j or 90°, so the equilibrium equation of the system

$$P = m\ddot{X} + c\dot{X} + kX$$

becomes

$$P\,\sin\,\omega t = (\,-\,mx\omega^2 + jc\omega x + kx)\,\sin\,(\omega t - \phi),$$

and we can represent the harmonic components of this as the receptance, the displacement in the mode direction divided by the cutting force

$$\frac{X^\sim}{P^\sim} = \frac{1}{k - m\omega^2 + cj\omega},$$

and recalling that the natural frequency is

$$\Omega = \sqrt{\frac{k}{m}}$$

and the damping coefficient is defined as

$$\zeta = \frac{c}{2\sqrt{mk}},$$

we can rewrite the receptance as

$$\frac{X^\sim}{P^\sim} = \frac{1}{k\left[1 - \dfrac{\omega^2}{\Omega^2} + 2\zeta j\,\dfrac{\omega}{\Omega}\right]}, \tag{3.4}$$

and this can be represented as real (G) and imaginary (H) components where the real component is

$$G = \frac{1 - \dfrac{\omega^2}{\Omega^2}}{k\left[\left(1 - \dfrac{\omega^2}{\Omega^2}\right)^2 + 4\zeta^2\,\dfrac{\omega^2}{\Omega^2}\right]}, \tag{3.5}$$

and the imaginary component as

$$H = \frac{-\,2\zeta\dfrac{\omega}{\Omega}}{k\left[\left(1 - \dfrac{\omega^2}{\Omega^2}\right)^2 + 4\zeta^2\,\dfrac{\omega^2}{\Omega^2}\right]}, \tag{3.6}$$

The receptance then can be written as

$$\frac{X^{\sim}}{P^{\sim}} = G + jH.$$

Now let us assume that the cutting force variation ΔP is described by an equation

$$\Delta P = br\,(Y_0 - Y), \tag{3.7}$$

where b is the width of cut, r is a cutting force coefficient depending on the cutting conditions, and by recalling equations (3.2) ($\Delta P' = f(\Delta P)$) and (3.3) ($Y = f(X)$) we can write

$$\frac{Y}{\Delta P} = \frac{X}{\Delta P'}\cos\alpha\,(\cos(\beta - \alpha)).$$

$\cos\alpha\,(\cos(\beta - \alpha))$ is a factor μ, representing how aligned the cutting force is to the forcing displacement and the machine mode direction:

$$-1 \leqslant \mu \leqslant 1.$$

Consider this for complete overlap, ($M = 1$), the case of plunge cutting

$$\frac{Y}{\Delta P} = \frac{X}{\Delta P} = G + jH \tag{3.8}$$

If the expression for cutting force (equation 3.7) is put into the equation it gives

$$br\,(Y_0 - Y) = \frac{Y}{G + jH} \tag{3.9}$$

The stability condition is

$$\left|\frac{Y}{Y_0}\right| = 1, \tag{3.10}$$

i.e. if Y, the output amplitude, is bigger than Y_0, the input amplitude, regenerative chatter will set in, resulting in an increasingly bad surface finish and noisy and unstable cutting conditions. So by rearranging (3.9):

$$br\,(G + jH)\left(\frac{Y_0 - 1}{Y}\right) = 1$$

$$\frac{Y_0}{Y} = 1 + \frac{1}{(G + jH)br}$$

$$\frac{Y_0}{Y} = \frac{(G + jH)br + 1}{(G + jH)br},$$

and using the stability condition (3.10)

$$\frac{Y_o}{Y} = 1 = \frac{[(brG + 1)^2 + (brH)^2]^{1/2}}{[(brG)^2 + (brH)^2]^{1/2}}$$

gives

$$(brG)^2 + (brH)^2 = (brG + 1)^2 + (brH)^2$$

and

$$2brG = -1.$$

Therefore the critical chip width b_{cr} is given by

$$b_{cr} = -\frac{1}{2rG}. \qquad (3.11)$$

For b_{cr} to have a positive value, G, the real component of the receptance, must be negative.

Plotting the single degree of freedom receptance as the sum of its real and imaginary components, Fig. 3.16 shows that the maximum value of G applicable is the negative value, G_{min}.

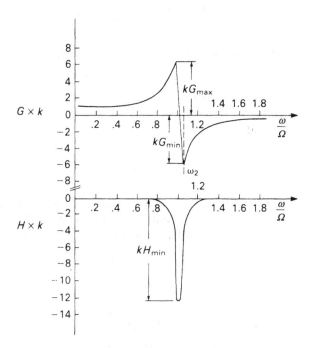

Fig. 3.16 The real and imaginary parts of receptance

$G_{\min}$ is given by

$$G_{\min} = -\frac{1}{4\zeta(1 + \zeta)k} \tag{3.12}$$

and the frequency, ω_2, by

$$\omega_2 = \Omega(1 + \zeta), \tag{3.13}$$

so that the critical chip width becomes

$$b_{cr} = \frac{4\zeta(1 + \zeta)k}{2r}. \tag{3.14}$$

In practice, the experimental value of limiting chip width, $b_{\lim}$, at which regenerative chatter sets in, is represented by curves shown in Fig. 3.17. The deviations are caused at low speed by changes in the direction of the cutting force P and damping in the cutting process itself. The lobing effect is also observed in experiments. This simple analysis has shown the importance of damping within machine tool structures to reduce vibration. It will now be clear why machine tools have been traditionally built from cast iron, which has high inherent damping due to the soft graphite flakes in its microstructure.

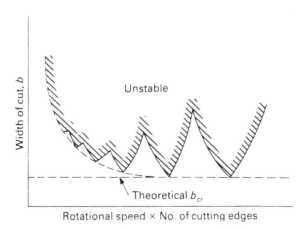

Fig. 3.17 A cutting stability plot

3.7 Programmable processes

In this chapter we have examined some of the problems associated with tailoring a process and machine to be programmable. If we require even more processes to be programmable there are equally complex problems associated with each process that must be solved before this is possible. Processes that are programmable include two-dimensional sheet metal nibbling and cutting, some assembly processes, and arc and spot welding.

Having addressed the programmable process, we now turn to the programmable linking of processes by examining the fundamentals of the robot manipulator.

As a closing comment we should say that the CNC machine tool has been one of the most powerful of the technologies applied in recent years, and can be applied in isolation as the solution to many problems that do not require the installation of more complex systems. The machining centre, when combined with the computer aided design (CAD) to CNC link (sometimes called computer aided part programming (CAPP)) has had a significant effect on many manufacturing processes. This has enabled the rapid and efficient production of dies and moulds for die casting, injection moulding and bulk and sheet metal forming processes with associated reductions in product lead time.

4 The robot – a handling device, a manipulator

After reading this chapter the reader should understand:

- the different types of robot manipulator;
- robot skeletons and geometries;
- a method of translating an end effector Cartesian position to a robot joint position;
- the tooling constraints on the application of robots.

4.1 Introduction

An industrial robot is a programmable machine that is used to transport objects around the manufacturing workspace. Figure 4.1 shows an industrial robot, the ASEA IRb2000, and Fig. 4.2 shows an AGV (automated guided vehicle), which is a robotic device that can only move in two horizontal directions. Figure 4.2(a) shows an AGV for transport and Fig. 4.2(b) shows an AGV being used as a travelling assembly station. Starting from a formal definition of the industrial robot, we will examine in this chapter the construction of robots and the problems associated with the manipulation of objects in three-dimensional space. We will close by discussing briefly some of the grippers that can be used to handle objects. Robots have been applied in industry for spot and arc welding, spray painting, assembly and metal cutting as well as the more familiar handling tasks.

4.2 A definition of the industrial robot

The ISO (International Standards Organization) definition of an industrial robot begins:

> An automatic servo-controlled reprogrammable multifunction manipulator having multiple axes, capable of handling materials, parts, tools or specialised devices through variable programmed operations for the performance of a variety of tasks.

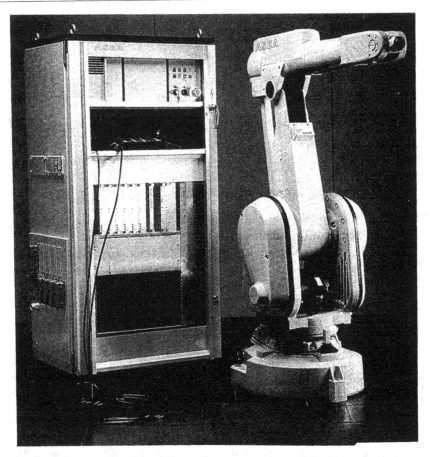

Fig. 4.1 The IRb2000 robot
Source: ASEA

4.3 Discussion of the definition

In this chapter we will confine ourselves to the discussion of the robot as a mechanical device. The key phrase in the definition above that really defines the robot as a machine is 'a . . . servo-controlled reprogrammable manipulator'. We have indicated the function of servo-control in the previous chapter and we will discuss robot programming in Chapter 6. This chapter will therefore concentrate on the manipulation aspects of the robot.

The definition emphasizes that robots are multifunction reprogrammable devices. This is rarely true. Many robot types are configured for specific application areas such as spray painting, which demands a six-axis robot that is easily programmed, and 'two-dimensional' assembly, demanding a four degree of freedom machine with high repeatability and carefully designed stiffness, such as the SCARA (Selective Compliance Arm for

(a)

(b)

Fig. 4.2 Wire guided AGVs (a) for transport and (b) for assembly
Source: Junheinrich

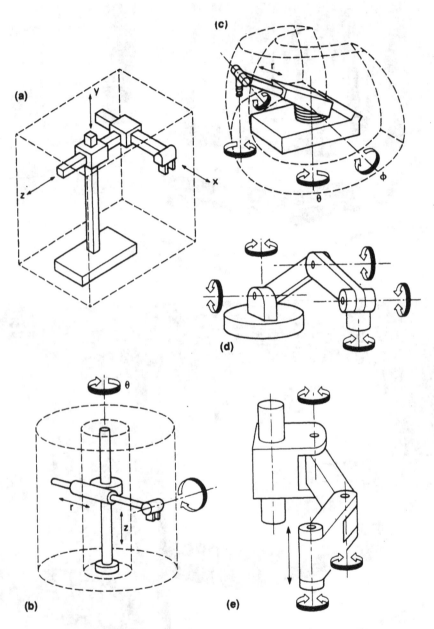

Fig. 4.3 The robot geometries: (a) Cartesian, (b) cylindrical, (c) polar, (d) jointed arm and (e) jointed arm, SCARA type
Source: British Robot Association

Robotic Assembly) robot. Robots are also rarely reprogrammed completely to carry out another task than that for which they were installed. This is usually because the non-programmable tooling required for a robot application is particularly expensive, and can only be justified financially for high production volumes. The life of the robot is therefore usually the product life. The rapid change of technology also rarely allows the old machine to be competitive with newer versions unless the particular new application is undemanding.

4.4 Robot types

Figure 4.3 shows the most commonly found robot geometries. The figure also shows the number of axes, usually the number of degrees of freedom that the machine has, and the robot workspace, the positions that the robot can reach. The geometries are:

(a) *Cartesian.* The axes of the robot are arranged in an *xyz* co-ordinate frame with a wrist, with three rotational axes, mounted on the last axis. Such machines can be made to be particularly accurate and are sometimes constructed as a gantry. Machines of this form are most frequently used for assembly. You will see from the figure that the robot workspace is a cube.

(b) *Cylindrical.* In this case the three major axes of the machine are arranged as an *rθz* co-ordinate frame. Such machines have a workspace described by a hollow cylinder and are usually less expensive than more sophisticated machines and are often used for simple handling applications.

(c) *Polar.* The early industrial robots were polar machines, with an *rθφ* co-ordinate frame. This geometry is now less common, because the volume of the available workspace consumed by the robot is large. The workspace is essentially a 'D' shape rotated through less than 360 degrees giving an area behind the machine that cannot be reached.

(d) *Jointed arm.* The most familiar robot configuration is that of the jointed or anthropomorphic arm. These mimic the shape of our own upper body and have waist, shoulder and elbow joints as their three major axes and a two- or three-axis wrist. Once again the workspace of these machines is a 'D' shape rotated through less than 360°.

(e) A special case of the jointed arm machine is the SCARA (Selective Compliance Arm for Robotic Assembly) robot which generally only allows joint rotations in a horizontal plane. (This machine will be discussed at length in Chapter 8.)

(f) *AGV.* The AGV, the automatic guided vehicle, is a robotic device that is only capable of programmable motion in two dimensions and is generally confined to following some form of guiding cable or painted line on a factory floor. Laser guided (Fig. 4.4), more

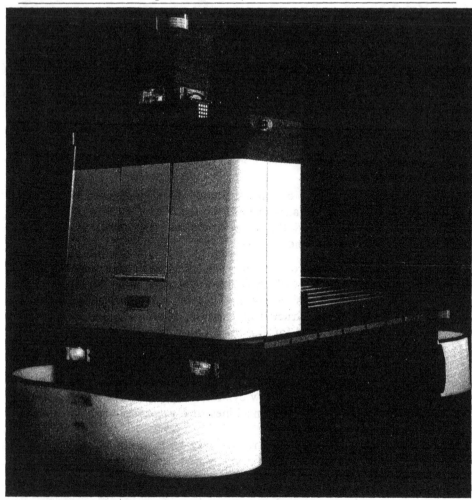

Fig. 4.4 A laser guided AGV
Source: GEC Electrical Projects

free ranging devices are beginning to be applied. In spite of its apparent limitations (when compared to the high degree of freedom manipulators) the AGV is widely applied to handle parts between cutting machines and welding stations in flexible systems, and to move partially completed assemblies between manual assembly stations in particularly flexible assembly systems.

4.5 Robot construction

The machines above are constructed from revolute (that is, rotating) and prismatic, sliding, joints as shown schematically in Fig. 4.5.

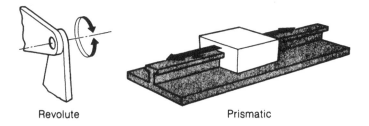

Revolute   Prismatic

Fig. 4.5 Robot joint geometries

It is necessary, if it is required to absolutely locate an object in space, that the machine has six degrees of freedom (corresponding with those of the object, see Fig. 4.6). These are usually the three major axes of the machine, as we have referred to them above, and a small three-axis wrist. Robot wrists are usually constructed from three revolute axes (Fig. 4.7), whatever the geometry of the remainder of the machine. These are often known as three roll wrists. It is not unusual to encounter five-axis machines for tasks that do not require the dexterity associated with six axes.

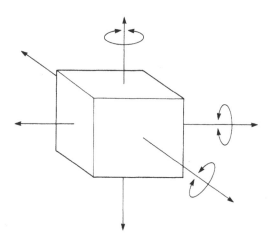

Fig. 4.6 The six degrees of freedom

All the machines described above are 'serial' robot arms in which the axes are arranged one after the other. An alternative (rarely encountered) approach is the 'parallel' axis robot, in which the axes are side by side. GEC have developed a number of these machines including the 'Gadfly', which has six parallel axes, and the 'Tetrabot' which has three parallel axes and a three-axis wrist. Figure 4.8 indicates the joint manipulations required for both types of machines moving from one point to another.

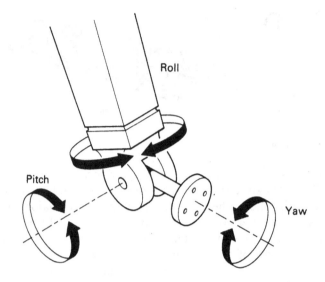

Fig. 4.7 A three revolute axis wrist

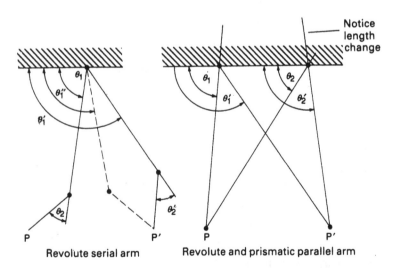

| Revolute serial arm | Revolute and prismatic parallel arm |

Fig. 4.8 Serial and parallel robot arms

Notice especially that the length of the links in the parallel machine change and that there are two positions for the serial arm at the target position. Six degrees of freedom are achieved in a parallel machine using the principle of the Stewart platform (the basis of the aircraft simulator) (Fig. 4.9) which has six parallel axes which can rotate and change in length.

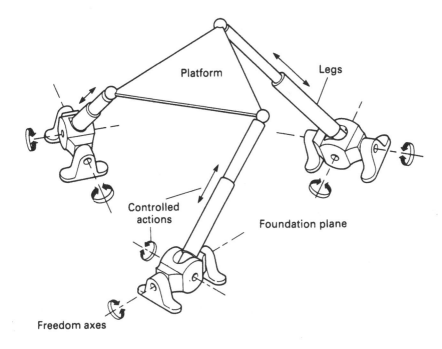

Fig. 4.9 The six degree of freedom Stewart platform

4.6 Manipulation with robots

One of the key issues to robotics is to consider how each of the individual joint positions chosen will move the end of a robot arm (often built up from revolute joints) to accomplish a manipulation in a three-dimensional Cartesian workspace. In this section we will examine the skeletons of robots and indicate one of the many solutions to the manipulation problem.

4.7 The robot skeleton

A robot can be regarded as a series of links that connect together the joints of the machine. Figure 4.10 shows a pair of joints joined by such a link. The figure indicates that the link length, a, is regarded as the shortest perpendicular distance between the joint axes, and that the twist angle of the joint, α, is the rotation, in a right hand screw sense, of the second joint with respect to the first. Figure 4.11 indicates the geometry of a joint showing the joint offset distance, d, the perpendicular distance between the links at the end of the joint, and the joint rotation, θ. To achieve manipulation in a revolute robot the joint rotation is varied and similarly

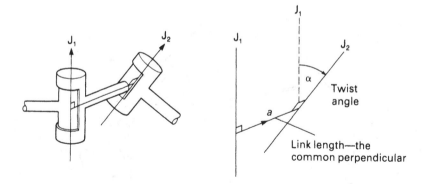

Fig. 4.10 Joints on a link

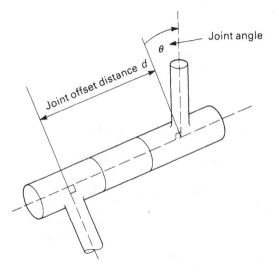

Fig. 4.11 The geometrical relationship between joints

in a prismatic machine the joint offset is varied – these are then known as the joint displacements.

Figure 4.12 shows the Unimation Puma robot, a small six-axis revolute robot, and Table 4.1 shows the geometric parameters of the same machine. The skeleton of the Puma robot is shown in Fig. 4.13. You will find that drawing the skeleton of a commercial machine from a schematic of the machine and the joint parameters helps to understand the characteristics of the machine. By convention, if two adjacent revolute joint axes are parallel then the common perpendicular (which embodies the joint offset of both joints) is chosen such that the joint offset distance at the first joint

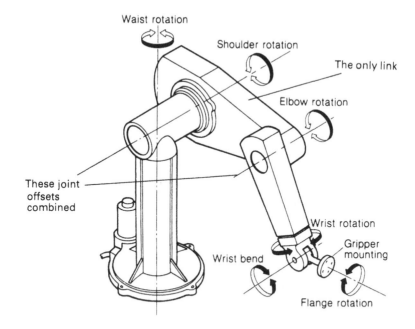

Fig. 4.12 The Puma robot
Source: Unimation

is zero. It will also be apparent that if the axes of joints intersect one another there will be a link of zero length between them – this is drawn on the skeleton as a dotted line.

Table 4.1 Geometric parameters of the Puma 560 robot

Joint number, j	Joint offsets, d_j(mm)	Joint rotation, θ_j(degrees)	Link length, a_j(mm)	Twist angle, α(degrees)
1	660	$-160 < \theta_1 < 160$	0	270
2	0	$-225 < \theta_2 < 45$	432	0
3	149.5	$-45 < \theta_3 < 225$	0	90
4	432	$-110 < \theta_4 < 170$	0	270
5	0	$-100 < \theta_5 < 100$	0	90
6	56.5	$-266 < \theta_6 < 266$	0	0

The geometry and kinematic properties are now completely described. These parameters can be used to determine the relationship between a reference frame fixed in the base of the robot and one in the gripper mounted on the flange of the machine. This kinematic description is particularly important as it enables the robot to be programmed off-line, by allowing the calculation of the joint movements required for relative motion, using CAD (computer aided design) systems which operate in Cartesian co-ordinates. This process is sometimes known as the

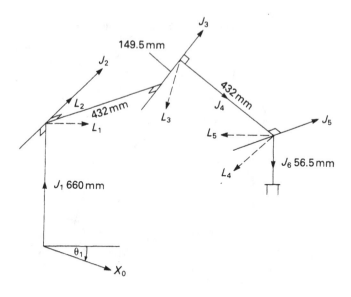

Fig. 4.13 The Puma skeleton
Source: Earl

'Cartesian-to-joint-space transformation'. The section below describes a particular algorithm (a step by step method), which is fast enough for real time implementation, devised for this process by Featherstone for a six-axis machine but is only presented here for a three-axis machine.

To indicate the complexity of this problem, consider the three degree of freedom manipulator shown in Fig. 4.14. You will see that the manipulator can be configured in four different ways to achieve the target point. For a six degree of freedom machine there can be 36 configurations. Many robots are configured so that they behave as a three degree of freedom manipulator, and are terminated by a wrist with the axes intersecting at the end point of the third axis. This simplifies the calculation process significantly. These four solutions are known as a left and right handed solution and an elbow-up and elbow-down solution. The different solutions can be useful as they allow the optimal configuration to be applied in particular circumstances (where, for example, parts of the robot workspace are occupied by tooling) and where it is required to apply high forces to the end-effector which must be resisted by the joint motor torque. Figure 4.14 also indicates how the moment on a joint can be minimized by selection of the machine configuration.

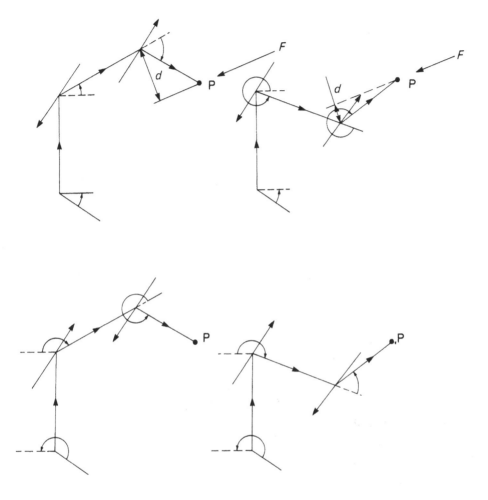

Fig. 4.14 The four ways of reaching P and optimizing motor torque $F.d$.

4.8 Cartesian-to-joint-space transformation*

Consider the robot shown in Fig. 4.15. The links l_1, l_2 and l_3 are co-planar and when $\theta_1 = 0$ they lie in the y-z plane of the co-ordinate system of the robot base.

To transform from the base to the end-effector co-ordinate system, the base co-ordinate system is rotated about the z axis by the angle θ_1 (that

* Material in section 4.8 is based on Featherstone (1983). Figure 4.15 is modified from that source by permission of MIT press.

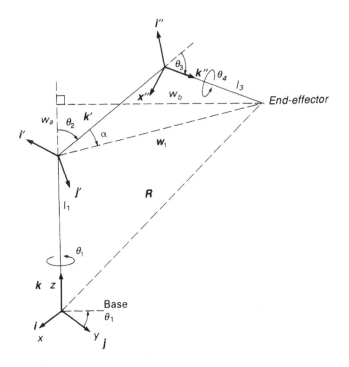

Fig. 4.15 A manipulator
Source: Featherstone (1983)

is rotating through $\theta_1 k$) and translating the origin along the z axis by a distance l_1 (that is moving through $l_1 k$), similarly through $-\theta_2 i'$ and $l_2 k'$, $-\theta_3 i''$ and $l_3 k''$ to reach the centre of the wrist, where i', j', k' and i'', j'', k'' are unit vectors in co-ordinate frames based in the second and third robot joints.

The position of the end-effector expressed in joint co-ordinates is the vector of the joint angles θ:

$$\theta = (\theta_1, \theta_2, \theta_3). \tag{4.1}$$

Similarly the position of the end-effector in Cartesian co-ordinates is the vector R where

$$R = (r_x, r_y, r_z) \tag{4.2}$$

The transformation problem is given R, find θ. The position of the end-effector from joint 2 is:

$$w_1 = R - l_1 k. \tag{4.3}$$

If the manipulator is reaching forward, i.e. θ_2 is positive, then

$$\theta_1 = \text{atan2}\,(-R_x, R_y), \tag{4.4}$$

where atan2 (x, y) is defined as $\arctan(x/y)$ for the four quadrants and

$$\theta_2 + \alpha > 0. \tag{4.5}$$

But if it is reaching backwards (θ_2 negative)

$$\theta_1 = \text{atan2}\,(R_x, -R_y) \tag{4.6}$$

and

$$\theta_2 + \alpha < 0 \tag{4.7}$$

If the cosine rule is applied to the triangle with the sides $l_2\ l_3\ w_1$ this gives

$$\cos \theta_3 = \frac{w_1^2 - l_2^2 - l_3^2}{2l_2l_3}. \tag{4.8}$$

The angles θ_2 and α are determined as follows:

$$\tan\,(\theta_2 + \alpha) = \frac{w_b}{w_a} = \frac{\sqrt{(w_{1x}^2 + w_{1y}^2)}}{w_{1z}}, \tag{4.9}$$

and

$$\alpha = \text{atan2}\,(l_3 \sin \theta_3,\ l_2 + l_3 \cos \theta_3). \tag{4.10}$$

The parameters $\theta_1,\ \theta_2$ and θ_3, which position the end-effector correctly, are therefore resolved. There are a number of particular circumstances that need to be taken into account in the calculation of exact solutions – the reader is recommended to refer to the original work for these.

For a real six-axis machine it is necessary to position the remainder of the joints to move the end-effector into the correct orientation. This can be achieved using spherical trigonometry, so solving all the position transformations. It also is necessary to use similar algorithms to carry out similar velocity transformations to allow complete robot control.

4.9 The practical application of robots

When applying robots there are two particular properties of the machine that must be examined: the machine payload and its accuracy and repeatability.

The robot payload is the maximum mass of material that the machine can manipulate without any decrease in its specified performance. This is significant in robot selection, as the workpiece weight that is being handled will immediately constrain the variety of machines that can be applied to carry out a particular task. The payloads of robots are small when compared to the size and apparent strength of the machines, for example 'man-sized' six-axis machines rarely have a payload greater than 20 kg.

Robots are, because of their construction, inherently less accurate than machine tools. This means that, when they are applied, the task to be carried out must be within the repeatability of the machine. The tooling around the robot must also then present parts to the robot in positions that vary less than the robot repeatability.

4.10 Grippers

When a robot is applied it is always necessary to tailor it to the task that it will perform by designing an end-effector or gripper.

We will close this chapter with a brief look at some gripper designs for handling operations. The most usually encountered gripper has two opposed fingers. These are usually pneumatically actuated and have parallel (where the gripper finger faces are always parallel) or purely pivoted actions. Parallel action in a gripper ensures that the object to be gripped is not displaced during the gripping process. The fingers can be designed so that components can be gripped internally or externally. A variety of gripper designs are shown in Fig. 4.16. Simple mechanics can be used to calculate the gripping force necessary for the gripper actuator to supply by considering object weight, likely object acceleration, geometry and friction coefficients between the gripper and object.

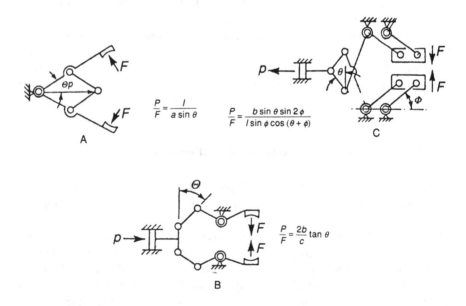

$$\frac{P}{F} = \frac{l}{a \sin \theta}$$

$$\frac{P}{F} = \frac{b \sin \theta \sin 2\phi}{l \sin \phi \cos (\theta + \phi)}$$

$$\frac{P}{F} = \frac{2b}{c} \tan \theta$$

Fig. 4.16 Robot gripper designs
Source: Chan (1982)

4.11 Sensory robots

This chapter has examined the robot solely as a manipulator. Devices such as robots are increasingly being integrated with binary sensor systems and high level sensory systems (such as force feedback) to behave in an intelligent manner – and hence to cope with more disorder in their environment. Sensors for manufacturing and the most well developed high level sensor, currently that of machine vision, are examined in the next chapter.

5   Sensors

After reading this chapter the reader should understand:

- the importance of simple sensors in manufacturing systems;
- touch trigger probes;
- the use of sensors to support unmanned machining;
- the essentials of machine vision;
- the use of complex sensors in manufacturing systems.

5.1 Sensors for manufacturing

There is a wide requirement for sensors in manufacturing automation. If we are to replace the handling, measurement and machine monitoring activities of a human operator we must necessarily go some way to replacing his eyes, ears and touch. This chapter will begin by examining simple sensors to determine position and monitor production processes, and then discuss vision sensing at some length. These technologies are an essential part of the integrated systems that we will discuss later in the book.

5.2 Sensors to monitor position

The use of vision to monitor the position of an object in the workspace can be expensive, technically risky and difficult to achieve with the necessary accuracy. It is usual to use as simple a sensor as possible to carry out a task. Simple position sensing tasks can be divided into two groups: those requiring high accuracy and those requiring lower accuracies. We will begin by looking at simple, low accuracy sensors usually known as proximity sensors.

5.3 Proximity sensors

The objects being carried around a manufacturing system are usually orientated and in a fairly fixed position. For example, items being fed into a manufacturing cell are likely to be deposited at the same point by a transport system. Their arrival at this point can be simply detected with a sensor that registers their proximity or nearness. The fact of their arrival can then be passed to the logical control system of the installation.

Proximity sensors are usually considered to be non-contact devices. However a similar function is carried out by the microswitch, which is operated by actual physical contact.

5.3.1 MICROSWITCHES

Microswitches are cheap and readily available. One of their main features is a toggle switch action, removing an element of uncertainty from the contact itself and transferring it to a mechanical toggle spring. This ensures that the logical state of the system is reflected by the fact that the switch contacts are either made or broken.

The following mechanical parameters can apply to microswitches:

1. *Pretravel* – the distance through which the actuator has to be moved before the switch operates;
2. *Differential* – the distance through which the actuator has to be moved back from the operating point to release the switch;
3. *Overtravel* – the distance that the actuator may be moved after operating the contact;
4. *Operating force* – usually expressed in terms of weight, the force required to operate the contact.

Powered mechanisms can never be brought to rest instantaneously and overtravel is seldom sufficient to accommodate the resulting displacement, so a roller or cam may be used to allow for unlimited overtravel without damage.

5.3.2 INDUCTIVE PROXIMITY SENSORS

The most frequently encountered true proximity sensor is the inductive type. As the sensor, an inductor, approaches a magnetic material and the magnetic circuit that includes the sensor is changed. This may alter the resonant frequency or the balance voltage of a sensing circuit; if the voltage then exceeds a pre-set threshold, the electrical output from the sensor is changed. This signals the logical presence of the magnetic material.

Such sensors (one is shown in Fig. 5.1) are usually adjusted with the threaded body to detect proximity at a suitable distance.

5.3.3 OTHER PROXIMITY SENSORS

Other less frequently encountered proximity sensors are:

1. optical devices, such as a light emitting diode (LED) and photocell

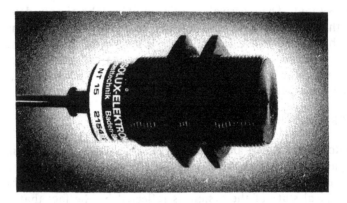

Fig. 5.1 An inductive proximity sensor
Source: Visolux

combination (the photocell views the LED and will change its output voltage when the LED is obscured hence detecting the presence of an object);
2. Hall effect devices, which once again show a change in output voltage in the presence of magnetic materials; and
3. capacitative devices, which change their capacitance in the presence of an object and can therefore detect it.

The optical and capacitative devices are more usually used with non-metallic materials.

5.4 Touch trigger probes

Touch trigger probes are used for a number of tasks requiring accurate position measurement. These activities are carried out on co-ordinate measuring machines and machining centres.

They are used on co-ordinate measuring machines for post-process gauging, that is checking finished dimensions of objects when removed from the machine that has produced them. Figure 5.2 shows a co-ordinate measuring machine (CMM). Such machines have a probe, usually mounted at the end of three Cartesian axes. The object to be measured is placed on the bed of the machine, which is usually a granite surface table, and measured by either moving the probe around the object manually or under programmed control. Such machines are capable of measuring objects and correcting the measurements for small changes in the orientation of the object.

Their use on machining centres includes the measurement and sensing of the following:

● *Absolute workpiece position.* By measuring the true orientation and

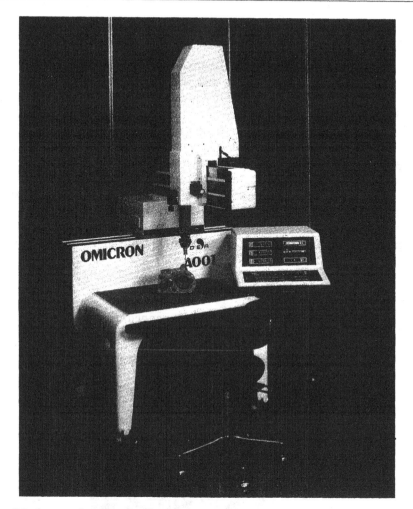

Fig. 5.2 A co-ordinate measuring machine
Source: DEA

dimensions of the workpiece it is possible to modify a general part program to accommodate variations in the position of the workpiece on its fixture.

- *Macroscopic workpiece sensing*. This confirms that the workpiece is present or has a hole where a hole is expected. This can be used to deduce which component is present on the machine and hence which part program is to be selected.
- *The absolute position* of tool tips can be measured to determine tool settings and tool offsets.
- *Tool breakage*. Tool wear cannot be sensed because of the small dimensional changes associated with flank wear. This wear may be

sufficient to cause unacceptable surface finish but still may not be detectable.

The touch trigger probe (Fig. 5.3) is a precision, omni-directional trigger device consisting of a probe body and a stylus. It may be mounted in a machine spindle or on a machine bed. Whenever the stylus is deflected in any direction, the probe sends a signal to the machine controller via a cable or other method and the current co-ordinates of the point of contact are known. After the contact has been removed the novel design of the stylus support allows the stylus to return to its datum position with a very high accuracy. An extremely small deflection (one micrometre) is required to trigger the probe. The stylus force necessary to cause this deflection is adjustable from as low as 20 g and the contact may occur at speeds up to 100 mm/s. An overtravel permits the stylus to deflect after contact, without damage.

Figure 5.4 shows some typical applications of a touch trigger probe.

5.5 Tool breakage sensing

One of the critical requirements for unmanned machining is broken tool detection. In the event of non-detection the machine tool will

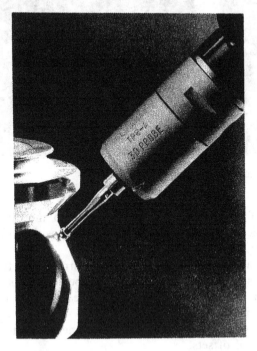

Fig. 5.3 A touch trigger probe
Source: Renishaw

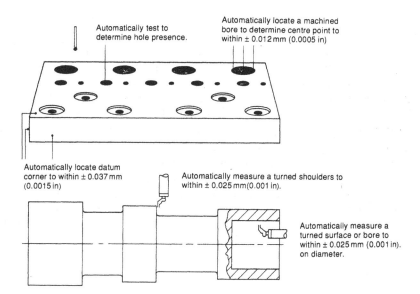

Fig. 5.4 Touch trigger probe applications
Source: Cincinnati Milacron

continue with its part program to either damage the machine tool or create scrap.

For minimally manned manufacturing, tool breakage is usually detected in three ways: before the process (pre-process), within the process or after the process (post-process). The tool size usually distinguishes between the approach taken.

- *For small tools*: Proximity sensors or touch trigger probes are usually used to detect the presence of small tools or the edges or inserts of larger tools. This assumes that if the tool is not at its specified position it is broken. This is a pre-process activity. The workpiece can also be inspected with the touch trigger probe. When there is no hole where one is expected, the tool generating it must have broken. This is a post-process activity.
- *For large tools*: There are many methods of detecting the cutting force on a large tool, and variations of this cutting force can indicate the condition of the cutting tool. The most common practically encountered methods are to measure the machine spindle motor current, voltage and speed to calculate almost instantaneous measurements of motor torque and power. These measurements allow some adaptive control of the machining process. Adaptive control implies that the cutting conditions themselves are modified (adapted) as a result of measurements made on the cutting process itself. Limits on the measurement allow tool breakage detection. This is discussed further below.

5.6 Torque and power monitoring

Cincinnati Milacron produce a system that allows torque controlled machining on prismatic machining centres. A microprocessor takes conditioned signals from the machine spindle drive, carries out computations on these, and communicates the results to the machine controller.

The operation of the device is shown in Fig. 5.5. The current, voltage and rotational speed of the spindle drive motor are sensed and are provided as digital inputs to the microprocessor. The 'firmware' (programs permanently residing in the microprocessor memory) uses these values to calculate the drive power and torque. An initial 'air cutting' or 'tare' torque is subtracted from the total torque to obtain the net cutting torque. The microprocessor then performs a series of comparisons with six pre-set limits in the NC program. If any of the six limits should be exceeded, remedial action can be initiated. The six limits are as follows.

5.6.1 ADAPTIVE CONTROL LIMIT

This permits efficient cutting using adaptive control. It ensures that the maximum net cutting torque felt by the cutter or drill never exceeds a set value. When cutting torques higher than 50% of the set limit are calculated, a signal is sent to the controller to reduce the feedrate. ,

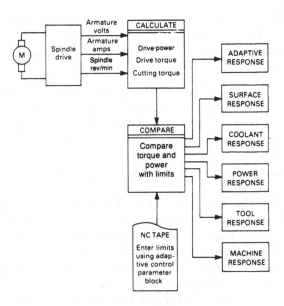

Fig. 5.5 Torque and power monitoring
Source: Cincinnati Milacron

5.6.2 SURFACE SENSING LIMIT

This is a very low current torque setting that is used to test for the surface of components. When a surface is contacted by the cutter the torque observed will rise and this will confirm the presence of the component.

5.6.3 COOLANT CONTROL LIMIT

This again is a very low torque limit. The response to a 'coolant on' command in the part program is delayed until the cutting torque exceeds the limit, so that coolant does not flow unless cutting is actually in progress.

5.6.4 MAXIMUM POWER OVERLOAD LIMIT

This is also an adaptive control limit which works on motor power rather than net torque. It starts to operate at the rated power and ensures that the total drive power does not rise to twice this figure.

5.6.5 CUTTING TOOL PROTECTION LIMIT

This is intended to protect the cutting tool. If the cutting torque should exceed the set limit, an interrupt occurs. The action taken then depends on the type of tool being used. For example, in the case of a tapping cycle, exceeding the limit would cause the tap to reverse and feed out of the hole.

5.6.6 MACHINE PROTECTION LIMIT

When the total drive torque exceeds this limit, axis movement is halted. This is an extremely high limit and is to protect the machine from irreversible damage by behaving as a 'software shear pin'. The default (unmodified) limit is based on the machine strength. This default limit can be lowered and could be set to provide, for example, protection to the workpiece based on its strength.

The monitor therefore provides a measure of machine protection, workpiece protection, tool protection, surface sensing capability, adaptive control and coolant control.

The monitor indirectly indicates tool wear and breakage if the modes of failure are severe enough to cause high values of torque or power to be observed by the system, since wear and breakage are not directly sensed. It may also indicate breakage by the absence of a torque in the surface sensing mode.

The microprocessor operates in a continual loop of reading spindle motor variables, calculating torques and power and comparing these values with the six limits. Even though the loop operates very quickly, there is about a 70 ms delay between the torque acting on a cutting tool and the computation of that torque by the microprocessor. There is a further delay of 10–100 ms between the time that the microprocessor issues an interrupt and the time that the machine response is observed. During this time cutting tools can move a significant distance.

Many other commercial systems solely monitor spindle power by measuring the voltage and current being consumed by the spindle motor. It is particularly difficult to detect all but the most severe conditions with this approach.

5.7 Force sensing

The sensing of force can also be used to adapt robot applications in processes such as assembly. Force sensing is usually carried out using strain gauge bridge-based load cells and is as yet, not widely applied in manufacturing.

5.8 Machine vision

The use of machine vision is one of the faster growing applications of programmable machines in industrial automation. This is due to the large effort put into the development of machine vision by the academic and industrial community. This effort reflects not only a desire to give machines an 'eye', but an increasing demand to monitor quality on the production line with a remote sensor and the need to accomodate by sensing, for example, parts in a variety of positions and distinguish between different types of parts. This has become necessary because of the disorder in the manufacturing workplace and demands to increase the variety of parts that pass through a production facility.

We can see that the use of machine vision is not confined to using the output of vision system to control an industrial robot, but can be used in a variety of tasks, e.g. in the control of quality of paint finishes and the inspection of satisfactory component placement and soldered joints in electronic printed circuit board manufacture.

The practical use of machine vision has emerged because of a change in the sort of problem that vision has been used to solve. Much of the early work in vision tackled large and very complex problems like the 'bin picking' problem. This uses a vision system to facilitate the selection by a robot of disorientated components from a pile in a bin or stillage. This is a testing task, which has been solved with varying degrees of success. The difficulties of such a general problem have led those applying the technologies to move to more easily resolved domains.

It is now practicable to use commercial machine vision systems in many areas, such as parts recognition, parts picking and the control of welding processes, provided the application area and technology is carefully chosen.

The remainder of this chapter outlines the steps to be considered in the application of machine vision and indicates the principles of the process. The machine vision process consists of five steps: image generation, image enhancement, image processing, image analysis and the use of the image information.

5.9 Image generation

When attempting to apply machine vision, it is imperative that the task is likely to yield to the machine vision process. Speculative vision tasks should not be tackled. A careful examination of the scene to be viewed will quickly resolve whether it is practicable or not.

The guidelines that should be applied to the scene are:

- the images of the objects can be represented as simple two-dimensional shapes; and
- the workspace should be arranged such that the workpieces do not intersect in the field of view of the camera.

It should also be remembered that machine vision is monochrome and not readily able to distinguish colours.

Unfortunately, in machine vision, the object needs to be viewed in high contrast lighting conditions that highlight the parts of the image to be analysed. The analysis and development of the lighting conditions can often be the most time-consuming part of developing vision applications. The recognition task must be independent of changes in ambient lighting conditions. Such changes include the on/off switching of fluorescent lights and variations in sunlight falling on the scene through windows. The process must be independent of changes in the surface properties of the workpiece, for example, reflectivity changes, the presence of rust spots and changes in the orientation of rolling marks in sheet metal.

One problem that may be encountered is that of a time lag between image capture and analysis. This can be associated with the high computational costs of examining a scene image point by image point. This problem is exacerbated by the use of more complex vision algorithms. Stroboscopic methods can be used, however, to stop moving images and make them easier to examine. There are also problems associated with 'blooming or comet tailing' of the image in the camera due to a very bright stationary or moving spot of light.

One of the most successful methods of lighting objects which do not require full field views is the use of structured light. The principle of 'structured' light is indicated in Fig. 5.6. A projector generates a narrow

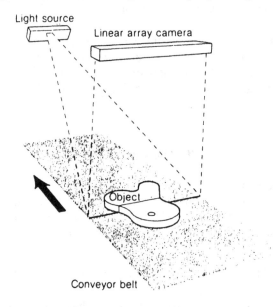

Fig. 5.6 The generation of structured light

band of very bright white light or sometimes laser light. When such a band of light falls on a plane surface the light is just observed as a strip of illumination. However, when an object is introduced into the light beam by a conveyor, for example, the strip of light becomes distorted to an observer not directly in the plane of the light beam, to give an image of a number of lines and arcs (Fig. 5.7). Such figures are dependent on the shape of the workpiece and the images produced by different shaped weldments and examples are shown in Fig. 5.8. The advantage of using white light, rather than monochromatic light, is that it is reflected more uniformly from a range of colours.

The field of view that has been created by the careful lighting of the workpiece must now be viewed by a camera. There are two types of camera used in commercially available vision systems, the first of these is the conventional tube type camera, the usual 'television' camera. Such cameras are often bulky and can be sensitive to shock, but are inexpensive. The other family of cameras are known as solid state charge coupled devices (CCDs), which change their electrical properties when exposed to light. These are still comparatively fragile and expensive but they are very

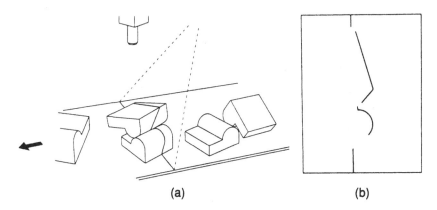

(a)   (b)

Fig. 5.7 Generation of structured light image: (a) line of intersection of a plane of light with objects on a conveyor; and (b) resultant image

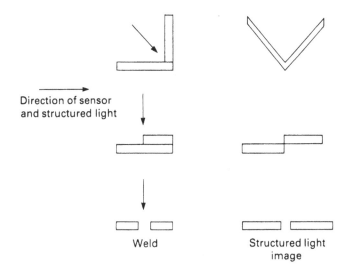

Fig. 5.8 Structured light images of weldments

compact. Charge coupled devices have the further advantage that they provide an image already composed of discrete elements. The light sensitive elements in the CCD device are arranged as an array as shown in Fig. 5.9.

The orientations and configurations of scene and single camera are varied, though two are of immediate and practical significance. In the first of these a single camera is fixed in a stationary position in the workplace so that it is viewing the scene of importance, and in the second the camera or a fibre optic image collection device is mounted on a machine axis arm or gripper; this is known as an 'eye in hand camera' and both arrangements are shown in Fig. 5.10. The eye in hand camera can go anywhere within

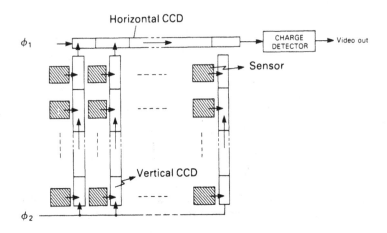

Fig. 5.9 A schematic CCD array

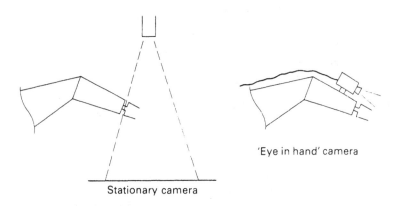

Fig. 5.10 Camera and manipulator arrangements

the scene to be examined, and can often allow higher resolution of the scene. The stationary camera, however, is much easier to install, does not get in the way of the task and is easier to calibrate.

Both sorts of camera are configured to produce a digitized image, made up of a matrix usually of up to 512 x 512 individual 'pixels'. A pixel (or less frequently pel) is a 'picture element'. Each pixel has a value with a range usually of up to 256 called its 'grey level'. The grey level of the pixel is proportional to the intensity of the image at the corresponding point. An example of such an image is shown in Fig. 5.11. Such an image can be stored as a two-dimensional array, with each element of the array having a value equal to the intensity at that point. Such images are stored in 'frame stores'.

Fig. 5.11 A grey level image

The discussion above has concentrated on the formation of single images from single cameras. Two cameras (Fig. 5.12) can be used to generate depth information using binocular stereoscopic vision. This has been used, for example, to determine the distance of automotive wheel arches from a sealant dispensing robot.

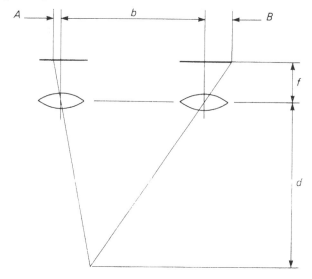

Fig. 5.12 Binocular stereoscopic vision

The distance, d, of the object from the lenses of the two cameras is given by

$$d = \frac{fb}{A + B},$$ (5.1)

where f is the distance from the image plane to the lens, b is the pitch of the cameras and A and B are the displacements of the image from the centrelines of the cameras.

Depth perception can also be achieved by using a single camera observing a scene at an angle and a reference image created by a target object in a known position. The distance of the new object from the camera, d, is related geometrically to the distance of the reference object, R, by the expression

$$d = \frac{R A}{B},$$ (5.2)

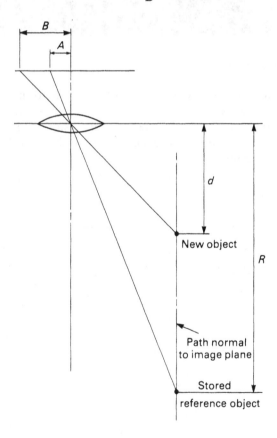

Fig. 5.13 Depth perception with a single camera

where A and B are defined as the distance from the reference image and the distance of the new image from the centreline of the image plane. The principle of this is shown in Fig. 5.13.

5.10 Image enhancement

Once the grey level image has been produced it must now be optimized for image processing. One of the most frequently encountered methods of image enhancement is to change the contrast of the picture by increasing its apparent contrast around the average grey level of the image.

Non-uniform lighting conditions can sometimes also be compensated for by using shading corrections. These can compensate for linear lighting variations across a scene by, for example, subtracting a linearly increasing grey level from the values of the grey levels measured at each pixel.

5.11 Image processing

We are now in a position to begin to examine and analyse our picture. There are two operations that are usually applied; binary imaging and edge detection of a grey level image.

First consider binary imaging. Binary imaging produces an image with only two intensities, black and white. A grey level image as generated above is converted into a binary image by selecting the range of grey levels it is required to observe by thresholding. Each pixel above a certain

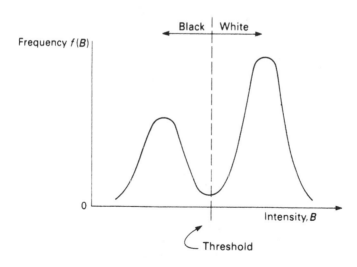

Fig. 5.14 Thresholding to create a binary image

selected intensity level (the upper threshold) is treated as white, and each below treated as black (Fig. 5.14). A lower threshold can also be set. These, by careful selection of the grey level thresholds or 'cut-offs', will give an outline view of an object in the workspace. Binary imaging has the problem that it cannot show any detail on the surface of an object and cannot distinguish between two intersecting parts.

The other method of image processing, edge detection, is more sophisticated and more difficult to use practically. To produce an edge detected image, as shown in Fig. 5.15, the grey level image is examined pixel by

Fig. 5.15 An edge detected image

pixel to detect the sharp changes in intensity gradient that are associated with edges on the object in the field of view, and then these sharp changes

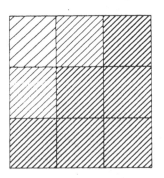

Fig. 5.16 A pixel and its eight nearest neighbours

are tracked to give the edges of an object. The edge detection process itself has to evaluate the local intensity gradient between the pixel being examined and each of its eight nearest neighbours (Fig. 5.16). This is essentially numerical differentiation, and can be done by using a matrix operator. Sobel is the most usually encountered operator, which works in the following manner.

Consider the pixel under examination as $g(x,y)$ and the surrounding pixels being numbered from 0 to 7. The small part of the image can be represented as a matrix where $B(i)$ is the value of brightness or intensity at each point i, the grey level

$$B(0)\ B(1)\ B(2)$$
$$B(7)\ g(\ )\ B(3)$$
$$B(6)\ B(5)\ B(4)$$

To determine the intensity gradient, the Sobel operator in the x direction is:

$$S_x = \begin{matrix} -1 & 0 & 1 \\ -2 & 0 & 2 \\ -1 & 0 & 1. \end{matrix}$$

You will see that this weights the pixels in the x direction. The gradient in the x direction is therefore

$$g(x) = (B(2) + 2B(3) + B(4)) - (B(0) + 2B(7) + B(6)). \tag{5.3}$$

Similarly for the y direction

$$S_y = \begin{matrix} 1 & 2 & 1 \\ 0 & 0 & 0 \\ -1 & -2 & -1, \end{matrix}$$

and

$$g(y) = (B(0) + 2B(1) + B(2)) - (B(6) + 2B(5) + B(4)) \tag{5.4}$$

are the intensity gradients in the x and y directions. The local intensity gradient is then given by

$$g(x,y) = \sqrt{(g(x)^2 + g(y)^2)}. \tag{5.5}$$

When the edge detected image has been established it is usual to generate a thresholded binary image of the edges to use for the further analysis.

When the image has been generated it is usual to have some command in the image analysis software to allow extraneous isolated single pixels to be rejected from the image.

5.12 Image analysis and object recognition

Now it is necessary from (in most cases) the binary image to carry out further processing. Most of the operations to do this are based on simple arithmetic processes such as calculating the area of an object, its perimeter, position of its centre of gravity and minimum second moment of area. This information is usually sufficient to determine the difference between objects, their position and orientation. Such parameters can be used to teach an object to the memory of a vision system.

Some objects may have similar values of some quantities and require more sophisticated techniques to be used to distinguish between them. Two such methods are boundary pattern analysis and template matching.

Boundary pattern analysis is carried out on an object whose centroid has been determined. The edge of the object is plotted as a polar plot around this centroid. By comparing the polar plots of taught objects the object can be identified independent of its orientation, and this orientation can be rapidly determined by the displacement of the plot from that of the

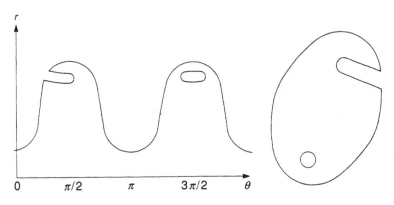

Fig. 5.17 A polar plot of an object
Source: Davies (1984)

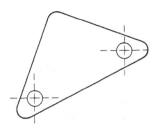

Fig. 5.18 Template matching using salient features
Source: Davies (1984)

taught object. Such a plot, and the object that generates it is shown in Fig. 5.17.

In *template matching* the vision system is taught an object to create a 'template', a complete record of the object. When a new image has been observed the template is moved around the new image until it matches. This is expensive (in terms of computer time) for a complete object because all positions and orientations of the object must be tried. This is therefore often carried out solely on the salient features of an object, such as holes in a specific orientation (see Fig. 5.18).

5.13 Use of the image information

Once the image has been captured it can be used in a number of ways. It can be used for quality control in association with hard reject/accept automation. It can also be used in operations where the image analysis is, for example, used to select a robot part program based on a critical dimension of an object. The image information can also be used to servo a robot arm in real-time. This is the hardest to achieve and to date the most rarely encountered in practice. Vision tasks that need to be fast are generally 'hard-wired' using logic circuits rather than being driven by software on more general purpose hardware. Figure 5.19 shows the image generated in an automatic inspection system when inspecting a printed circuit board.

5.14 An example of vision linked with a robot

One of the most significant applications of vision and robots was pioneered by VS Engineering. This is the insertion of windscreens and rear windows to automotive body shells. It is necessary to control the robot with vision information because of the wide variation in the windscreen aperture size and position. This is because the body shell is an assembly of a large number of components of relatively low stiffness.

The installation, shown in Fig. 5.20, is built on two levels. In the upper level the first stage of the process is carried out, involving the careful and accurate application, by robot, of a chemical primer and special adhesive to the windscreen rim which is used to seal and secure the glass to the body shell. This glass is then transferred to the glazing area by a vertical conveyor.

On the final assembly track the body shell is held stationary for a few seconds by an intermittent transfer mechanism while the glazing robot withdraws the glass from the vertical conveyor. The robot gripper incorporates four cameras and structured light sources, the glass being held

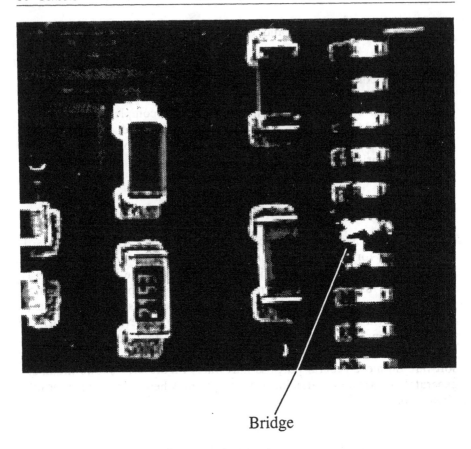

Bridge

Fig. 5.19 Image generated in an automatic inspection system

by vacuum cups. The robot approaches the aperture in the body, and stops 30 mm from the aperture. The vision system then observes the exact position of the aperture and a supervisory computer takes this position data and uses it to drive stepper motors in the gripper which positions the glass correctly. The robot then carries out the assembly operation.

Systems of this type are now widely applied in the automotive industry worldwide. Newer assembly systems use robots that are capable of accepting offsets from external sensor systems. This means that the grippers applied in the final assembly system do not require their own actuation systems and the end effector is therefore lighter and more compact. This, in turn, means that the robot itself can be smaller. Some companies only use robot systems to lay the adhesive bead on to the windscreen. This screen is then manually assembled to the car body. This creates a space-efficient installation that does not control the speed of the final assembly track or constrain the variety that can be sent down it.

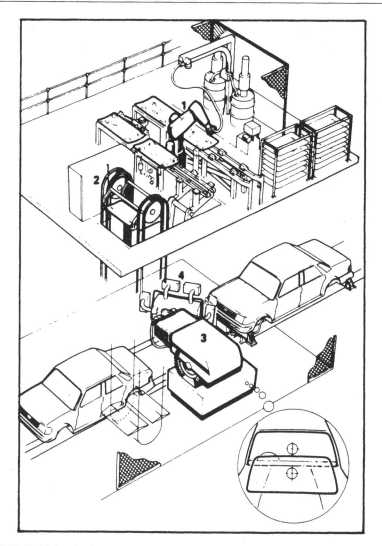

Fig. 5.20 Vehicle glazing
Source: ASEA Manufacturing Systems

5.15 Colour vision

It is tempting to consider the application of colour vision techniques in manufacturing, particularly in applications such as food processing. It should be noted that colour vision operates using three different features, typically hue, saturation and intensity, in contrast to the single grey level used in monochrome imaging. This can make the matching process 3^3 times harder. Hue, H, saturation, S and intensity I are determined from an RGB (Red, R, Green, G, Blue, B) colour camera signal as follows:

$$I = \frac{(R + G + B)}{3},$$

$$S = 1 - \frac{3 \min (R, G, B)}{R + G + B},$$

and

$$H = \frac{(G - B)}{3 (R + G - 2B)} \quad \text{if } \min (R,G,B) = B$$

$$= \frac{(B - R)}{3 (G + B - 2R)} + \frac{1}{3} \quad \text{if } \min (R,G,B) = R$$

$$= \frac{(R - G)}{3 (R + B - 2G)} + \frac{2}{3} \quad \text{if } \min (R,G,B) = G$$

6 Software for single machines

After reading this chapter the reader should understand:

- methods of programming machine tools and robots to achieve a position in space;
- manual part programming;
- the essentials of CADCAM;
- the place of CADCAM in CIM.

6.1 Introduction

In this chapter we are going to look at software for the control of individual programmable machines. Individual machines, such as machine tools, programmable placement machines and robots are usually devices made up of a number of servo-controlled, linked axes with a local microprocessor in control of each axis and another microprocessor controlling the interactions between axes, and perhaps the outside world.

The controller of an 'ideal' programmable machine (Fig. 6.1) can be viewed to have three separate elements: closed loop servo-control for the machine axes, 'sequence control' to handle the interactions of the machine with its environment and its internal non-servo control activities, and a controller to process complex sensor information. The sequence control

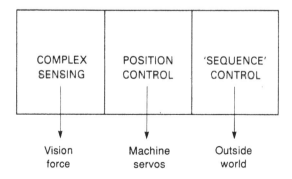

Fig. 6.1 An 'ideal' machine controller

element (an activity similar to that carried out by programmable logic controllers) will be examined more closely in the discussion of the control of cells and systems later in the book.

The goal of a machine tool or robot is to move a cutter or a robot end-effector in space at the correct time in the machine sequence in response to a program command to create or move the object. We are now going to look at the software and programming methods used for such devices.

6.2 Machine tool programming

NC (numerical control) part programming is the process of obtaining from the detail drawing of a part, control data which can be used by a numerically controlled machine to produce the part.

To be able to program the machine tool to move in space, the programmer must have a co-ordinate frame to represent the axes, and rotations about the axes, of the machine. It is usual to use a 'right hand' convention for the relationship of these axes as shown in Fig. 6.2. The z axis is usually the axis of the machine spindle and increasing the displacement in the z direction increases the clearance between the cutter and the part being machined. The x axis is usually parallel to the work holding surface and usually horizontal. Conventions of spindle and chuck rotation (for lathes) are that positive rotations (anticlockwise when viewed from positive z) cause a right hand screw to recede from the workpiece, i.e. position rotations do not apply cut. Cutting with tools with a conventional right hand helix therefore usually requires a negative rotation.

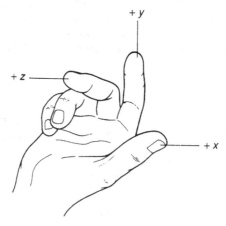

Fig. 6.2 The right hand axis convention

6.3 Manual part programming – machine tool code programming

Each line of a program is referred to as a BLOCK which is divided into WORDS, each word consists of a word ADDRESS and each block is terminated by a line feed/carriage return.

An example of such a line is shown in Fig. 6.3 together with a simple English language translation. To identify the elements of the line, the sequence number word just identifies each block, and these are arranged

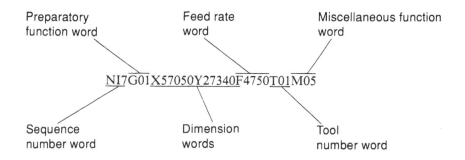

Fig. 6.3 A program block. The block, with sequence number 17 says, 'Do a linear interpolation (G01) from where you are to a point 57.05 mm away on the x axis and 27.34 mm away on the y axis at a feed rate of 4.75 mm/min with tool number 1 and the spindle on (M05).'

in ascending order. The dimension words (see Fig. 6.4 to help in the definition) specify the location of the tool centre point. The addresses available are X,Y,Z (the main linear motions of the machine), U,V,W (subsidiary linear motions on extra slideways), A,B,C (rotary motions) and I,J,K (for the generation of circular arcs).

The feed rate word, the element of the line that defines the cutting feed, is either programmed directly (for example 5000 mm/min gives F5000000) or as a feed rate number FRN (calculated as velocity over distance). R, rapid translational motions can, for example, be specified to move the tool to above the workpiece to speed up the machine cycle time.

The preparatory function words or 'G codes' are not dimensional, and have mainly to be executed before the dimensional instructions that actually define the position of the tool. G codes define the manner of the motion, for example G01, linear interpolation, tells the machine to go in a straight line from where it is at present to the new point specified in the block. Other G codes are:

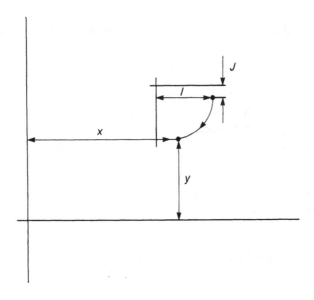

Fig. 6.4 The definition of dimension words

G00   Positioning mode of control (non-synchro-ed drives)
G01   Linear interpolation
G02   Circular interpolation

.

.

.

G81   Canned Cycles (a sequence of predetermined motions)
—
G89

G90   Absolute mode (target position measured from the machine zero)
G91   Incremental mode (target position measured from present position)
G92   Defines tool position

Some are MODAL and apply until cancelled and some are NON-MODAL and only apply to the block in which they appear. The miscellaneous function words or 'M codes' are essentially machine sequence control functions and allow the non-servo activities of the machine to be carried out. Examples are:

M00   Program halt (stops spindle and coolant – restarted by cycle start)
M03   Spindle on clockwise
M04   Spindle on anticlockwise
M05   Spindle stop

.

.

M08 Flood coolant on
M09 Coolant off

These instructions are either put into the machine manually by typing directly at its controller (known as manual data input), or are prepared as a punched or magnetic tape and read by a tape reader incorporated in the machine controller. The loading of these instructions by DNC systems (direct numerical control) is indicated in Chapter 9.

6.4 Computer-assisted part programming

Manual part programming is very time-consuming. The tape generated has to be proved on the machine and each machine has a different subset or slight variations in the definitions of the effects due to the codes. Computer-aided techniques have therefore been developed. Although still manually extracting the data from an engineering drawing, the programmer, at a terminal remote from the machine, specifies the motions required of the tool in a higher level (more 'English-like') language than the machine tool codes indicated above. These devices running the high level languages (sometimes known as 'processors') require the use of 'post-processors' to convert high level code generated into machine tool code instructions. Each particular type of machine tool and controller will have its own particular post-processor because of its particular geometry and the particular actions and sub-set of machine tool codes implemented on that machine.

The most frequently encountered high level programming language or processor used is APT (Automatically Programmed Tools). It is a very large and complex language and many simpler sub-sets have been developed for more specialized applications such as ADAPT, for 2.5 axis contour milling and EXAPTI (Drilling), II (Turning) and III (Milling). In EXAPT, tools, feeds, speeds and machining sequences can be automatically generated by the 'processor'.

Two essential elements are required for CAPP (computer assisted part programming*); a description of the geometry of the part and the tool

---

* The reader should be aware that the acroynm CAPP is ambiguous, it is sometimes taken to mean computer aided process planning. This definition can imply that a computer system is used to specify and organize the sequence of different manufacturing processes carried out on different machine tools, for example, prepare blank → rough machine → heat treat → finish machine. Computer systems exist that support the carrying out of this process. The use of computer aided part programming here emphasizes that the process is used to generate the software that is used to control the metal cutting machine tool and that the process also uses considerable knowledge of the geometry of the part to carry out its task.

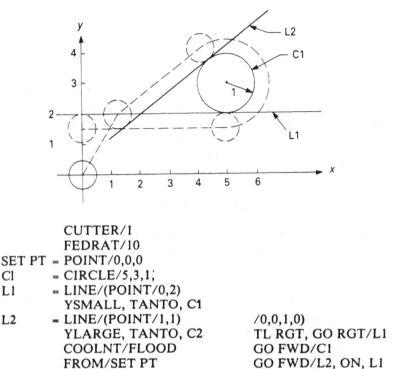

```
                CUTTER/1
                FEDRAT/10
SET PT  =  POINT/0,0,0
Cl      =  CIRCLE/5,3,1,
Ll      =  LINE/(POINT/0,2)
           YSMALL, TANTO, C1
L2      =  LINE/(POINT/1,1)          /0,0,1,0)
           YLARGE, TANTO, C2         TL RGT, GO RGT/L1
           COOLNT/FLOOD              GO FWD/C1
           FROM/SET PT               GO FWD/L2, ON, L1
           INDIR                     GO TO/SET PT
           P/(POINT/0,2,0)           COOLNT/OFF
           GO L1 (PLANE              STOP
```

(from a set point (0,0,0) define a circle radius 1 at (5,3), draw two lines at tangents to this passing through (0,2) and (1,1), turn the coolant on, passing through the indirect point (0,2,0) follow the outlines of these lines on the outside of the figure, the tool turning with a right hand screw and on the right hand side of the first line, with a 1 diameter cutter with a feedrate of 10, go back to set point, turn the coolant off, stop)

Fig. 6.5 An example of APT

motion relative to this geometry. The toolpath in APT is defined in terms of a three-dimensional tool moving along a pair of intersecting surfaces until 'movement' is stopped by a third surface. An APT statement has a major part to the left of a slash, which shows the motion or logical part of the information in the instruction, and the other which shows the numerical data. An example of a short APT program to generate a shape is shown in Fig. 6.5.

It is usual to simulate the output of such programs using suitable graphics software, to ensure that the path generated by the APT program actually generates the shape of the part and that the tool motions are possible and collision free.

6.5 Conversational part programming

Machine tools are being marketed which allow conversational part programming; they are aimed at the small workshop where it is usual for
the machine operator to program the machine. It is generally found on
machines that do not generate very complex geometries, such as CNC
lathes and three-axis milling machines. Conversational part programming
essentially consists of a simple graphics screen supported by menu driven
software. The operator is guided by the software to input the material and
geometry of the part via a keypad, and extracts this data manually from
an engineering drawing. The software further guides the operator to input
the tooling available, its dimensions and the feeds and speeds that are
applicable to the material of the part. The controller will then generate
the tool path, and simulate it on the graphics display. Figure 6.6 shows a
series of screens, showing blank and part geometry input and tool path
generation.

6.6 Programming from a CAD system, CADCAM

Machine tool part programs can be generated directly from CAD descriptions of the part. This is what is usually known as CADCAM, the link
between computer aided design and computer aided manufacture.

The most basic form of CAD is interactive computer aided two-
dimensional (2D) drafting, which uses interactive computer aids to speed
the drawing process. This is often said to be automation in the form of a
'drawing-processor'; compare this with a word-processor.

By capturing the engineering drawing there is already a great deal of
data representing the form of our product in a computer. It would be
sensible to use this data for other tasks within the organization requiring
product data rather than generating this data again. At the CAD screen,
computer assisted part programming (CAPP) can be used to prepare code
often via APT for metal cutting machine tools. The prepared code can be
transferred to the machine by a data network.

6.7 Representation of (3D) three-dimensional shapes

Perhaps the most significant part of the data stored in the database is the
data describing the product itself, sometimes called the product model. So
far we have only indicated the use of two-dimensional data, as two ortho-
graphic views, however three-dimensional representations are increasingly
being used as representations of the object and as the starting point for

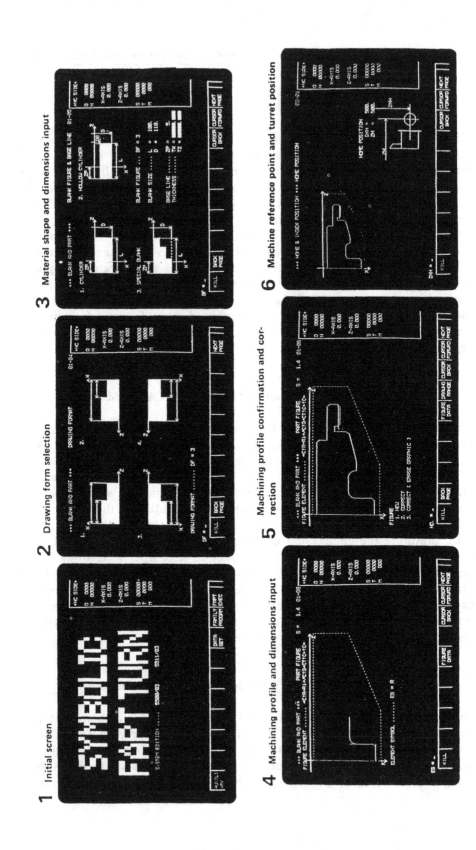

1 Initial screen

2 Drawing form selection

3 Material shape and dimensions input

4 Machining profile and dimensions input

5 Machining profile confirmation and correction

6 Machine reference point and turret position

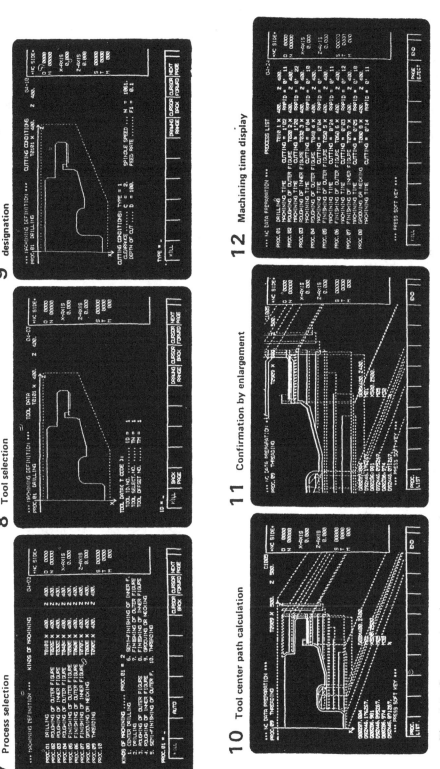

Fig. 6.6 Conversational part programming screens
Source: Fanuc

control program generation both for NC and robotics. The section indicates the methods of representation of this 3D data.

A 3D computer model of an object can be represented inside a computer memory. The model, when built, can generally be viewed from any direction and more conventional 2D data extracted from it. There are a number of significant types of modeller.

6.7.1 WIREFRAME MODELS

A wireframe model of an object is made up of a set of three-dimensional co-ordinates which define the end points of lines in space. The lines can be curved or straight. This type of model, because it is solely described by its edges and vertices, can only provide partial information about the shape of an object.

(a)

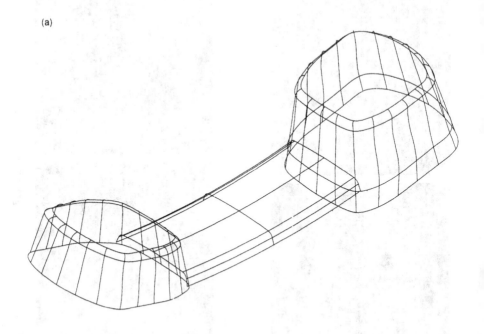

6.7.2 SURFACE MODELS

Surface models give the description of an object in terms of points, edges and faces between edges. Essentially, the models produced are in the form of a mesh, constructed from a set of measured or calculated co-ordinates. The co-ordinate points which describe the form of the surface of the object are input to the modeller which uses this geometric data to create a meshed surface. Figure 6.7(a) shows such a model.

This form of model is frequently used to generate NC part programs of objects with particularly complex surface profiles, such as moulds and dies, and turbine blades. Figure 6.7(b) shows tool paths generated from the model.

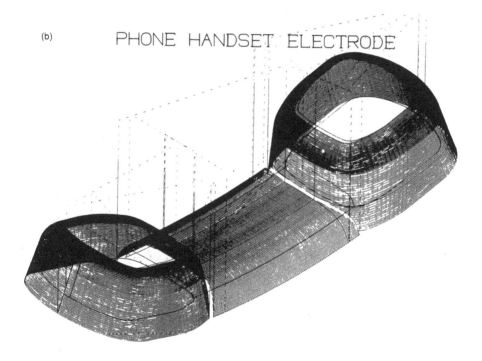

(b)

PHONE HANDSET ELECTRODE

Fig. 6.7 (a) A surface model and (b) associated cutter paths
Source: NC Graphics

(a)

(b)

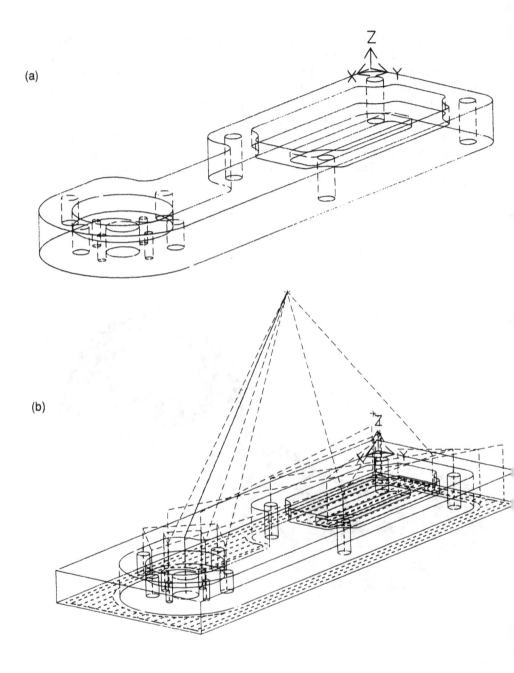

Fig. 6.8 (a) A solid model and (b) associated cutter paths
Source: BUILD Group

6.7.3 SOLID MODELS

A solid modeller is the most powerful of the three-dimensional modelling techniques as it provides the user with complete information about the outline, surface, volume and mass properties of the object. A solid modeller holds a complete description of an object. The description can indicate whether any point is inside or outside the object and whether it lies on the surface of any object. Solid modellers build up models from three-dimensional shapes that can be described mathematically. One of the methods, known as constructive solid geometry (CSG) uses Boolean operations on simple geometric primitives like blocks, cylinders and spheres. Figure 6.8 shows a component generated using a 3D solid modeller and tool paths generated using the model.

6.8 CADCAM in CIM (computer integrated manufacturing)

In the present atmosphere of international competition it is necessary to have a complete business strategy for the whole company. This implies that the company activity is integrated, using either people or the computer. Many pages could be spent arguing which hybird is the most appropriate for particular companies. However, while talking about program generation and the use of data, it is worthwhile discussing the CIM philosophies. Figure 6.9 indicates the usual CIM acronyms.

There are essentially two views of the meaning of CIM. One is that CIM provides solely computer assistance to all engineering and business functions and thereby increases the efficiency of the business. The other, by actually combining all the computer based sub-systems and tools in the manufacturing business, into the same large system, ensures that all sub-systems within the CIM system use the same data which is held in the same single database.

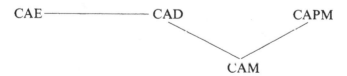

CAD   — computer aided design
CAE   — computer aided engineering (complex calculation aids)
CAPM — computer aided production management
CAM   — computer aided manufacture.

Fig. 6.9 The CIM acronyms

This single database contains static, once per design (CAE (computer aided engineering), CAD, CAPP and BOM (bill of materials) etc.) data and dynamic, once per manufacturing cycle (CAPM (computer aided production management), manufacturing schedules) data. This has a number of effects:

1. Data is not duplicated, the original design data describing the shape of the component is used to generate the part program.
2. There is no conflict of data, because a single current data source is used.
3. Data storage is minimized.

As a caveat, the systems that the descriptions of CIM and CADCAM in the literature imply rarely exist, and where the programmable manufacturing technologies exist they are rarely linked to the balance of the CIM systems. The only real links are generally those associated with CADCAM. This itself is usually the generation of NC program data from a 2D CAD system and transfer to the machine by DNC link. It is, however, a step towards the paperless factory. That such complete integration does not exist is perhaps fortunate. As world competition increases and product life reduces it is becoming recognized that rigid CIM systems may not be able to evolve sufficiently rapidly to accommodate the changes required by the

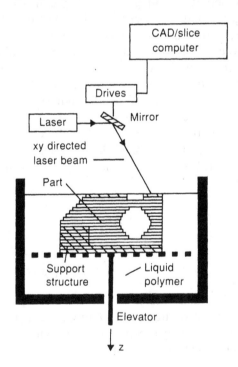

Fig. 6.10 Production of prototype components by stereolithography

manufacturing organization in today's dynamic environment. CADCAM systems, however, remain particularly important: newer generations of solid model based systems allow rapid visualization of potential product designs and rapid translation of designs into hardware – this shortens lead times and supports concurrent engineering practices.

Rapid prototyping methods have matured since the publication of the first edition of this book. The most well established of these, stereolithography, allows the production of prototype components in resin materials as is shown in Fig. 6.10. A laser beam is driven by a post processor working on CAD data. The beam cures layers of liquid photopolymer to build up the shape of the product. These components improve the ability of designers to visualize their products and have the potential to shorten the lead time required to generate product prototypes.

6.9 Robot programming

There are a number of methods of programming robots. They are presented below in order of increasing complexity, the stages in the discussion reflect the stages in development in robot system.

6.9.1 TEACH-BY-SHOWING

The most frequently encountered method of programming robots is called teach-by-showing or teach-by-lead (less frequently known as guiding). In

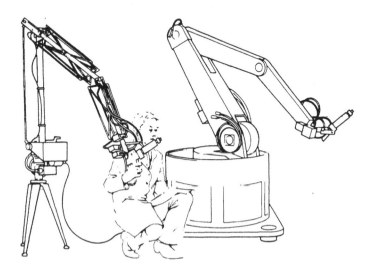

Fig. 6.11 A teach arm

this the robot is guided around the task workspace by the programmer, and its position recorded at intervals either automatically by the robot controller or by the operator programmer.

Let us describe the two methods in more detail. In paint spraying, for example, (a continuous path process) the robot is manipulated by the programmer actually holding the machine or a light weight 'teach arm' (Fig. 6.11) and the machine controller automatically records path points at intervals. Such a method is used for paint spraying because the process depends critically on the manipulative skill of the operator.

In the second method the robot is once again moved around the workspace, but in this case the robot is under power and is manipulated by the operator using a teach box or pendant (Fig. 6.12), which allows him to drive individual axes of the machine. The operator now records points in the workspace as he thinks fit by pressing a record button.

When the program has been prepared it is then replayed by the programmer so that the robot will pass through the taught program points to within the limits of the machine repeatability.

Such teaching methods, although simple and easy to master, do not allow for the programming of conditional actions or use of any of the editing facilities found on a more general purpose computer.

6.9.2 HIGH LEVEL LANGUAGES

The applications of industrial robots have become more demanding. Associated with these increased demands have been developments in robot programming languages that have allowed the machine to tackle more demanding tasks. These languages, as well as allowing pure motion control, permit the robot to control grippers, items in its environment and take data from other devices such as sensors. This section of the book will briefly indicate the elements of robot languages, while Chapter 11 will look more closely at the 'flow of control' sequence or logical aspects of robot languages.

There are essentially two forms of robot programming language: implicit or task level programming languages and explicit or robot level languages.

(a) TASK LEVEL PROGRAMMING

Implicit or task level programs, in which the robot is not instructed explicitly, step-by-step, what to do, are at present solely implemented at a research level and then only to carry out simple tasks.

In task level programming, the task, the job that the robot is required to carry out, is described in a very abstract way. An instruction in a task level program essentially represents a description of a state transition (the discrete step of changing from an initial state to the next state) in the

Manual Control Pendant

2 line LCD display,
40 characters per line

5 programmable
'soft' keys

'State' LEDs for
use in manual mode

5 function keys

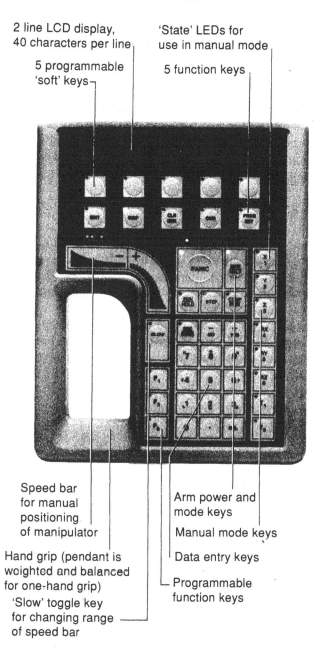

Speed bar
for manual
positioning
of manipulator

Hand grip (pendant is
weighted and balanced
for one-hand grip)

'Slow' toggle key
for changing range
of speed bar

Arm power and
mode keys

Manual mode keys

Data entry keys

Programmable
function keys

Fig. 6.12 A teach box or pendant
Source: Meta Machines

world of the robot. An example of such a statement for an assembly task (from Lozano-Perez (1983)) is

PLACE BEARING1 SO (SHAFT FITS BEARING1.HOLE) AND (BEARING1.BOTTOM AGAINST SHAFT.LIP)

(put bearing one on the shaft so that the shaft goes in the bearing hole and the lip on the shaft touches the bearing base).

This statement is a description of the initial state of the robot and of the desired state to be reached once the first state has been achieved. It is assumed that the robot controller has the necessary hardware and software to allow the high level instruction to generate the low level instructions to perform the actual state transition.

(b) EXPLICIT OR ROBOT LEVEL PROGRAMMING

Explicit or robot level languages are necessary to develop today's applications. They are used on the more sophisticated machines currently available commercially. Most of these languages have been developed by computer scientists and therefore look very like computer languages such as BASIC and PASCAL.

Such languages are continually and rapidly developing. One of the first commercial languages was the Unimation Language VAL, which shows the style of motion control used in robot languages. An example of a VAL listing from Unimation is

1. TYPE SAMPLE INSTRUCTIONS
2. REMARK THIS IS NOT A REAL PROGRAM
3. SPEED 30.00 ALWAYS
4. DRIVE 1,30,100
5. DELAY 5
6. DRAW 100.00,200.00,300.00
7. MOVES POINT1
8. OPENI

After displaying SAMPLE INSTRUCTION on the screen, carry out the following instructions at a speed of 30 units, drive Joint 1 through 30 degrees at 100% of the programmed speed, wait 5 s, draw a vector 100 mm in the x direction, 200 mm y and 300 mm z, then move to the taught Point 1 in a straight line, open gripper immediately.
(Notice commenting with the REMARK statement.)

This language has modal speed commands: recall the definitions in NC programs. The essential difference between robot and machine tool programming is a result of the fewer degrees of freedom and higher accuracy of machine tools. The robot usually follows a complex path from start point to end point, this path being generated by all the robot joints together. The machine tool follows simpler paths and the axes do not always co-operate so closely. In spite of this, commands exist that allow individual joints to be driven on their own and draw vectors in space with the robot end-effector. There are some restricted flow of control or sequencing commands available which depend on the input from simple sensors. In most robot level programming languages, a program is written as a sub-routine which is accessible from all other sub-routines.

There are in the order of 80 explicit robot languages in the research laboratories and in the market place, and some of the more well known of these are RAIL (Automatix), HELP (DEA) and AML (IBM). They all have very similar features to those outlined above. Later languages have many more sophisticated features that are specifically intended for cell and application control. These are examined in Chapter 11. One of the most recently written languages, KAREL has been specially designed to allow not only more sophisticated robot programming by a software engineer but to allow him to prepare special purpose application specific software which hides the complexity of the programming task, from the robot operator. KAREL is discussed in Chapter 11.

Such languages have the potential to allow some of the code necessary to program the machine not to be developed in the manipulator itself. This activity is referred to as offline programming. Offline programming is understood to have two elements, the development of the flow of control code and the development of the position instructions.

6.9.3 PROGRAMMING FROM CAD SYSTEMS

As one would expect, there has been a significant research effort to generate explicit robot programs for motion control from CAD descriptions of the particular robot tasks using the same philosophy as NC programming from CAD. These are generated using robot cell modellers, which will be discussed at length in the chapter on cells.

6.10 Flow of control

In this chapter we have hardly addressed any of the 'flow of control' commands in robot languages or PLC capabilities in machine tools. These are also known as sequence or logic control activities and essentially respond to the inputs from simple sensors. We will examine these in the chapter of the book on system software, where we will see how they can be used to control simple systems.

7 The manufacturing cell – the building block of systems

After reading this chapter the reader should understand:

- the definition of a manufacturing cell;
- the place of automated manufacturing cells in the manufacturing system;
- prismatic and revolute machining cells and their key differences;
- robot processing cells;
- geometric and logical simulation of cell operation.

7.1 Introduction

In the remainder of the book we will examine the integration of single machines into cells, into assembly systems (together with the implications of this for the design process), and into manufacturing systems. We begin by examining the manufacturing cell which is regarded by many as the building block of larger systems.

7.2 Cells of factory automation

There are many ways of considering the cell of factory automation. It can be viewed in an almost biological way as the smallest autonomous unit capable of sustained production. As the cell is automated the cell must be minimally manned.

The cell can also be viewed in terms of the items that it usually contains. In this way we can see it as a small collection of machines, which are closely co-operating with each other. Close co-operation can include, for example, the sharing of dimensional data between a measuring machine and a machine tool, and the sharing of a workspace between a robot and a turning machine. Such machines often work in parallel, some of the machines in the cell carrying out manufacturing tasks at the same time. Examples of this could include a robot picking up an unmachined billet while the machine continues to machine the previously loaded billet. A usual feature of such cells is that they are controlled by a supervisory computer, and this computer may be resident in one of the machines in the cell. Figure 7.1 shows an example of a multimachine turning cell at the Fanuc injection moulding machine plant in Japan.

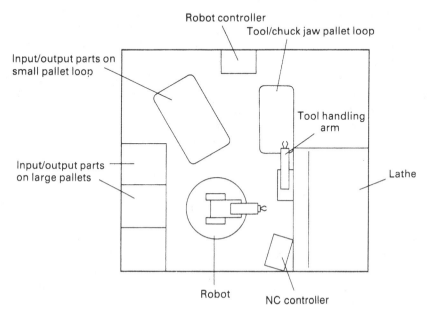

Fig. 7.1 A manufacturing cell
Source: Fanuc

Both of these definitions take an automated view of cellular manufacturing. It must be remembered that the cellular concept arose with the invention of group technology (GT) which was not necessarily automated.

Group technology considers that the cell arises from the physical division of machinery in the manufacturing facility into production cells. Each of these cells is designed to produce a parts family. A parts family can be considered as a set of parts that require similar machinery, tooling, machine operations and jigs and fixtures. A cell is usually used to transfer raw materials into finished products. In a GT cell the parts transport activity is usually carried out by an operator.

All of these concepts are transferable to more automated or minimally manned factory automation cells. The range of parts that each cell produces is, however, usually less than that in many conventional GT cells, and there are frequently less machines in the cell. This has often arisen from a necessity to minimize the mechanical and control complexity of the automated installation. In cells of factory automation, the cell has usually been built to get the optimal performance from a very expensive machine tool, and the balance of the machinery serves this particular machine so that it has minimum idle time.

A cell is the next level of integration from single machines, and characteristically consists of a number of machines of different types supplied by different manufacturers. Now consider our factory automation requirement for autonomy. Autonomy can be defined as minimally manned manufacture for a sustained period. It seems unlikely in the foreseeable future that the completely unmanned factory is practical, desirable or financially justifiable. A human operator will always be required for system patrol, maintenance and system recovery in some circumstances. The definition of minimally manned in the context of this book is the amount of manning consistent with economic and reliable system operation.

This autonomy requires that we have: 1. automated or programmable processes, 2. automated handling to and from these processes, 3. automated quality control or inspection of the performance of the processes together with rapid feedback of any necessary changes to each process, and 4. supervisory control and sensors to monitor and detect the cell condition and to decide the next activity. Many practical cells do not have the integral inspection facility, as this is carried out as in more conventional facilities by a patrolling operator. In factory automation it is usual to take good account of the capability of the manufacturing process and to design cells with a 'right-first-time' attitude. The increased adoption of statistical process control (SPC), however, means that inspection machines are, once again, increasingly included in cells.

Some single complex machines, such as prismatic machining centres and programmable electronic insertion machines, can almost be regarded as 'cells' in their own right, as they have been specifically constructed with integrated handling systems to run autonomously for long periods.

There is some confusion as to the definition of the expression 'cell' and 'workstation'. In some of the work coming from USA (particularly from the AMRF of the National Institute of Standards and Technology), a 'cell' is sometimes known as a 'workstation', and the workshop level of integration that is usually called a 'system' in the UK is known as a 'cell'.

7.3 Advantages and disadvantages of cellular approaches

The primary advantage of the cell in building systems for factory automation is that it reduces system mechanical and organizational complexity. This reduction permits easier financial justification, less technological risk, easier installation and commissioning – and, of course, allows stand alone islands of automation to be built. It is not clear whether the cellular approach reduces the complexity of the overall system control problem, although it is now clearer that the autonomy implied by automated and non-automated cellular factory structure supports people-based integration. Considering the cell alone, the advantages (when compared to conventional manufacture) are 1. reduction of the necessary 'organizational' control involved in bringing materials to a single cell rather than distributed machines, 2. reduced handling of parts, 3. reduced set-up time, often by using dedicated jigs and fixtures, 4. reduction of work-in-progress and inventory by careful attention to buffer design, and 5. a reduction in the need to expedite (progress chase) between machines.

The basic disadvantage of the cell is that of the island of automation. Even the most efficient automated cell, if it is isolated, may just move the manufacturing problem somewhere else in the complete manufacturing facility. A significant and frequently encountered example of this is the buildup of work-in-progress before or after the cell as this will not increase overall system efficiency. An isolated cell must therefore be a part of an overall manufacturing strategy. At a more mundane level any cell built must be capable of interfacing mechanically and electrically (power and control) with any further automation that is likely to be installed.

7.4 Applications of the cellular philosophy

We will now look at four short case studies in cellular manufacturing: a prismatic machining cell, a turning cell, a robot welding cell and a cell with automatic changeover. There are many other cellular applications, especially in sheet metal working, spot welding, fabrication and finishing. Most of the robot installations in industry that are not performing spot welding are manufacturing cells.

7.5 A prismatic machining cell

In the cell shown in plan in Fig. 7.2, two machining centres have been installed side by side and face a number of pallet stands which are at the front of the cell.

A pallet is essentially a portable, general purpose device that can be repeatedly positioned in a machine tool to secure a component during the cutting operation, but can then be removed simply and automatically from the machine. Components themselves are set-up and secured to the pallet rather than a machine table. It is usual for a pallet to be identified by a machine readable code so that it can be recognized as it is transported around the cell.

A simple railcar transporter sometimes known as an RGV (rail guided vehicle) is mounted on rails between the machines and pallet stands. The

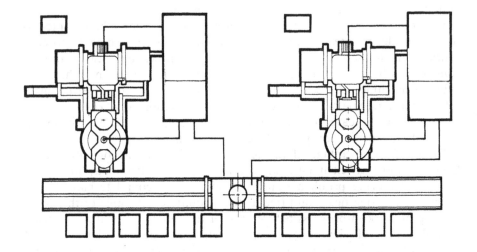

Fig. 7.2 A two machine prismatic machining cell
Source: KTM

entire installation is enclosed by perimeter guarding, but there is provision for access to two pallet stands for work loading and unloading. These stands have duplex tables which can be brought to a position for loading and unloading workpieces or pallets, and then moved to another point to be picked up by the transporter.

In a fully automatic sequence, the transporter picks up a pallet from one of the stands, carries it to a machining centre and discharges it on to a two-station rotary table. These operations are completed while the machining centre is in operation cutting the previously loaded component. At the end of a cycle, the rotary table indexes to bring the fresh work to the machining position and transfer the pallet carrying completed components to a point to await collection by the transporter.

Each machining centre is equipped with two 40-tool drum-type magazines, giving a total capacity for storing 160 cutters. This gives a margin over normal requirements, to enable 'sister' tooling to be accommodated. Sister tools are duplicates of either fragile tools or tools that wear rapidly. They are used to replace tools that have become worn or broken. There is provision for worn tools to be replaced with sister cutters after a preset number of operations have been completed.

For broken tool detection, each machine is equipped with an inductive probe built into a table-mounted arm, which automatically swings clear after a reading has been taken. This probe will detect variations in tool length, to detect that a drill has broken.

The machine also has an inductive coil at the spindle nose to enable signals to be received from a touch trigger probe. This can be loaded from the magazine with the toolchanger to be used for in-cycle component inspection. This facility is currently used on a limited scale mainly for probing drilled holes, to ensure that they are present, before tapping is carried out.

The cell was originally designed to produce 23 different types of parts and these can now be produced economically in smaller batches, avoiding the long production runs that were needed with earlier methods.

In addition, components are produced complete within the cell, though several set-ups are sometimes required. By avoiding the use of multiple machines, scrap resulting from setting errors has been eliminated. The fixtures used enable four or more components to be machined completely during one production run. Quality standards have been improved, and lead times have been reduced from between 20 and 24 weeks to as low as five or six weeks. Work-in-progress has been reduced by as much as 90%. The flexible production cell can now be operated for five, 24 hour days a week. Operators work with the machines during day and night shifts, but it has been proved that the cell can run successfully unmanned between 4.20 pm (the end of the day shift) and 9.00 pm (the start of the night shift).

Cells like this are beginning to be widely applied in both large and small companies, and are intended to be extendable by the addition of further machines and pallet stands. Figure 7.3 shows a similar cell with four machining centres. Such cells are also intended to be integrated into more complex systems, including additional similar cells and turning and assembly cells.

7.6 A turning cell

The turning cell, to make rotational parts, can be viewed as having the elements shown in Fig. 7.4: the turning machine itself, facilities for tool and workplace loading and unloading, and tool and workpiece gauging.

These can be integrated in a number of ways, as is indicated in Figs. 7.1 and 7.7. Figure 7.1 shows the use of different types of discrete machines; the cell uses an anthropomorphic robot to load two sizes of parts into the lathe, large parts being held on a pallet and small parts being held on a pallet loop. Cutting tools and different shaped chuck jaws are loaded into the lathe from another pallet loop by a simple lathe mounted robot as is shown in Fig. 7.5. A pallet loop is a continuous conveyor of discrete pallets. Figure 7.6 shows a group of similar cells configured into a large system and served by AGVs.

A rather more integrated and more dedicated turning cell is shown in Fig. 7.7. This is a turning machine that has been extended into a cell by the addition of input and output parts storage at the left hand side of the machine. Raw bar and finished parts are held in nests in racked pallets.

Fig. 7.3 A prismatic machining cell
Source KTM

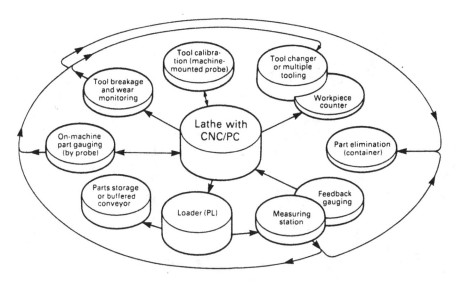

Fig. 7.4 The elements of a turning cell
Source: Georg Fisher

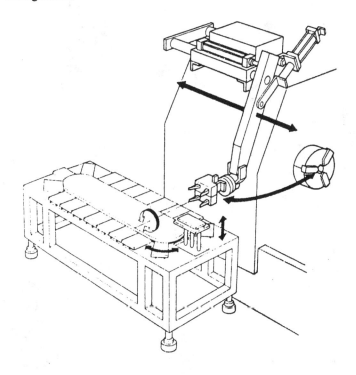

Fig. 7.5 Horizontal pallet loop and lathe mounted manipulator
Source: Fanuc

Fig. 7.6 Turning cells in a system

Fig. 7.7 A turning cell
Source: Georg Fisher

These are handled into the machine by an NC gantry loader operating over the top of the machine. A second loader also feeds tooling from the right hand end of the machine into a tool turret. This tooling is held in a drum magazine. The gripper fingers of the parts handling system also include a gauging station, to permit post-process parts measurement and error feedback of new tool offsets as tools wear or are replaced.

The tool offset modifies the nominal NC part program for the exact tool dimensions associated with a particular individual tool and its toolholder.

7.7 Differences between prismatic and rotational cells

There are a number of differences apparent when comparing each of the machining cells described for turning and prismatic machining. This is essentially because of the cycle time differences between turning and milling; turning cycle times are of the order of 5 min whereas milling times are about 30 min.

Cutting feeds and speeds are also higher in turning, as the process only uses a single point tool so that tool consumption is higher. In prismatic machining there is a significant problem associated with the management of the large number of cutting tools needed. There are fewer different types of tools used in turning as complex surface geometry can be generated by the programmable single point tool path. Careful design for manufacture and selection of the group of parts on a machine system can reduce the tool management problem.

These lead to an emphasis in rapid workpiece and tool changing on simpler machines in turning, with large buffer stores of input parts. The corresponding emphases in prismatic cells are the management of the large number of different tools required and the maintenance of workpiece to pallet accuracy.

7.8 A robot welding cell

All programmable cell-based automation is not solely based on metal cutting. There are many examples of the use of robots handling parts between a variety of machines, for metal forming and polymer processing and for assembly. Figure 7.8 shows a shell mould preparation cell in which a wax pattern is dipped into a series of stucco pots and hung out on to overhead conveyors for intermediate drying operations. We will now look at a robot-based cell or workstation where the robot is acting as a programmable processing machine rather than a handling device.

The particular task being performed is the CO_2 gas welding of hinges on to car doors. The workstation, part of a line (Fig. 7.9), is based on

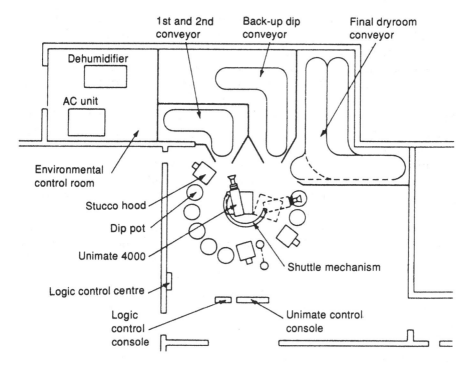

Fig. 7.8 A robot handling application

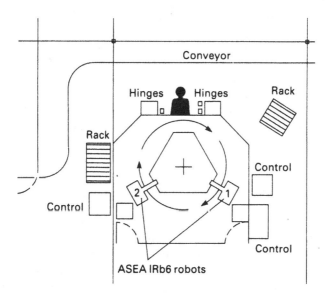

Fig. 7.9 A welding cell
Source: Rover Group

two ASEA IRb6 robots, each manipulating a welding torch. This torch is fed with welding wire and welding current from a welding controller, positioned at the rear of each machine. An operator loads partially completed doors (taken from a conveyor) and the upper and lower hinges into a welding jig mounted on a rotary table. The use of a rotary table is typical for a single robot installation. The rotary table contains three jigs, one at each robot and one at the loading station. One robot welds the upper hinge and the other the lower hinge. The operator removes and inspects completed doors.

Notice that the operator is carrying out the very complex tasks of part location, orientation and positioning in the jig, and inspection. The operator will also not load defective parts. It is very difficult and expensive to replace these functions with automation.

As the robot is actually carrying out the manufacturing process the workpieces presented to the robot must be within the robot welding process capability – which is a weld gap of not more than 0.5 mm – and the variation of the position of the hinge elements and the tip of the welding wire must be within a 0.5 mm sphere. This places high demands on previous processes, including the wire feed mechanism, to achieve and maintain wire straightness.

One of the other operating problems with robot welding applications is that the welding torch must be frequently cleaned to remove spatter, which prevents correct wire feed and causes variability in the welding process. This is accomplished by using airblast or anti-spatter oil cleaning, and mechanical brushing. Air blast cleaning can give porosity problems in the final weldment.

The whole installation is guarded and interlocked, so that it will not start until the enclosure door is closed and two start buttons are pressed (one for each of the operator's hands). The whole system has performance figures as shown in Table 7.1. You will notice the extremely high welding time of the robot, and also that each of the cycle times of the separate activities in the cell (the two robots and the operator) cannot be exactly matched.

7.9 A cell with automatic changeover

In early applications, cells were designed to produce a single part or, at the other extreme, to produce in very small batches. It was recognized that there was a considerable middle ground, where the requirement was a cell that was apparently dedicated to making a single part but could be rapidly changed over or reconfigured to make another part. This philosophy is apparent in some of the turning cells described above – it is however worthwhile describing a cell where those ideas have been applied explicitly.

Table 7.1 Robot workstation cycle times (s) *Source*: Rover Group

Robot cycle time		Lower Hinge Robot 1	Upper Hinge Robot 2
Arc time	27 (75% total)		22 (73% total)
Total time	36		30
Torch cleaning (8 s per 5 cycles)		1.6	
Table rotation		8	
Total		45.6	
Operator cycle time			
Load hinges and door		15	
Collect hinges and press start		6	
Collect door from conveyor		19	
Unload door and inspect		9	
Transport door to output buffer		9	
Total		58	

(Note that the operator and robot cycle times are not exactly matched.)

Such a cell, shown in plan in Fig. 7.10 consists of 1. a prismatic machining centre, 2. an industrial robot, 3. input and output magazines for aluminium extruded blanks, 4. a balance arm to assist in heavy lifting operations, and 5. a table holding fixtures.

The cell manufactures sixty variants of heat sink for thyristors. Machining is carried out by the machining centre, which has a pallet changer, tool storage magazine and automatic tool monitoring.

On each machine pallet within the cell there is a hydraulic clamping fixture which can hold a set of particular product variants. These can be re-tooled by an operator. When a batch is to be changed over the robot exchanges the machine pallet with one held on the fixture table using the balance arm to assist in the lifting of the heavy pallets.

The parts themselves are placed on standard input pallets, with a spacer between each layer of parts. This input pallet is then placed on the input conveyor which has room for 20–40 h of production, depending on the size of the part.

To load the machine, the robot with a photo-sensing gripper searches for the first component in each layer of the pallet and loads it into the machine. The gripper can pick up all the varieties of components machined by the cell. The gripper can also unload spacers and used pallets out of the cell.

When a new batch is to be made, the operator loads the data describing the batch into the system via a minicomputer keyboard. This data includes the number of pallets per batch, the robot and NC program identifications and the correct fixture location on the fixture table.

The installation can run unmanned for three shifts, and is expected to

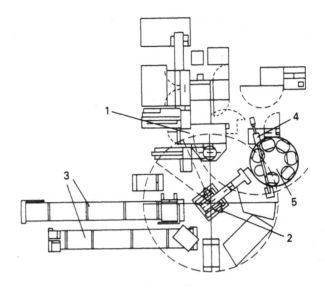

1. Machine tool; 2. Industrial robot;
3. Input and output magazines for cut aluminium
shapes placed in layers on pallets;
4. Balance arm; 5. Fixture table

Fig. 7.10 A machining cell with automatic batch changeover
Source: ASEA

produce in three shifts what two manually attended machining centres could produce in four shifts. It is also hoped that order fluctuations can be accommodated by running the cell over the weekend, the operator only having to attend the cell once or twice during this time. It is possible to produce small batches to reduce inventory and lead time.

7.10 Cell simulation

The design task for the manufacturing cell is one of machine selection and geometric layout, together with the development of a control logic to co-ordinate the interactions of machines in the cell.

Simulators have been developed from CAD systems which can simulate the motion of the elements of manufacturing cells moving in space. They are usually intended for the modelling and evaluation of industrial robot work places and tackle the geometric design problem. Figure 7.11 shows the output of such a simulator.

These CAD simulators allow the designer to build a static model of the workspace and cell components to check their spatial layout, ensuring that all the elements of the cell fit together and into the appropriate space.

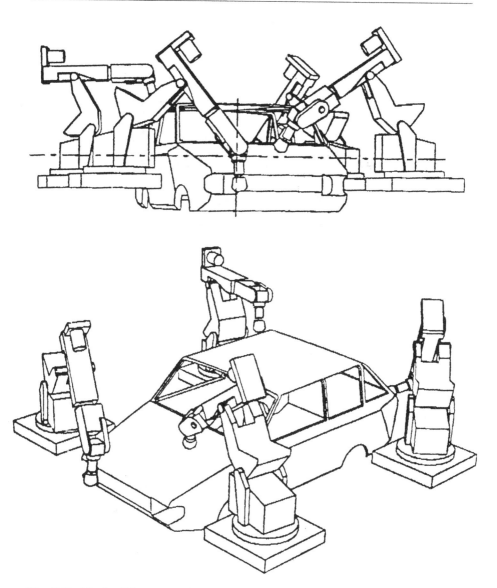

Fig. 7.11 Workcell layout model
Source: ASEA

The designer can also check that the cell elements are within the reach of the robot.

Such simulators are usually based on simple body modelling packages and represent the robot using simple shapes or primitives.

After a static simulation has been drawn, it can be animated to give a 'dynamic simulation' of the operation of the cell. In this dynamic

```
                                    T35761 ROBOT
                         CUR%  MAX%  CARTESIAN
                          87    87    -14.752
                          53    64    -55.656
                          20    78     -6.907
                          63    63    179.687
                           0     6   -130.370
                          54    54     47.483


                                  ->T3576  ROBOT
                         CUR%  MAX%  CARTESIAN
                          25    70     38.501
                          56    92      7.952
                          73    84    -18.131
                          30    93    134.853
                          52    52    -60.768
                          64    64     -5.177
```

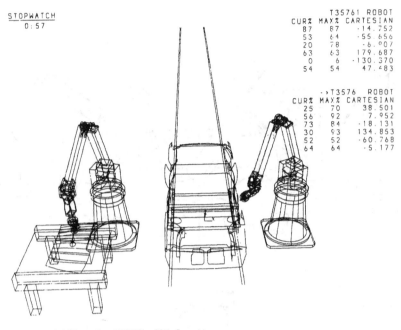

```
   : CALL   CAR4PO :   1 : CAR4POD : CALL Executing :   :   :
CALL_SEQUENCE: CAR4POA,CAR4POB,CAR4POC;
;**** APPLY SEALANT AND POSITION WINDSHIELD
CALL_SEQUENCE: CAR4POD,CAR4POE;
CAR4POD  [  8]:> GOTO_TPOINT: GLUESHD,TP30,NOP;

CAR4POE  [  9]:> GOTO_TPOINT: WSHSHD,TP1,NOP;
```

```
                                  ->T35761 ROBOT
                         CUR%  MAX%  CARTESIAN
                           7    90     75.583
                           3    81     -2.644
                          34    78     -7.150
                          80    89    147.691
                          37    76     90.011
                          74    77      0.007


                                    T3576  ROBOT
                         CUR%  MAX%  CARTESIAN
                          67    70      9.974
                          64    99    -59.844
                           7    98     10.735
                           8    93    179.589
                           7    71    -86.986
                           8    87     -3.284
```

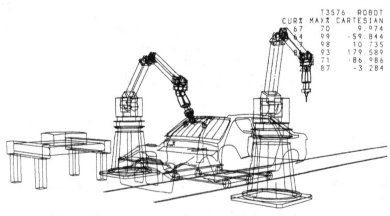

```
   : CALL : CAR4PO :   1 : CAR4POD : CALL Executing :   :   :
CALL_SEQUENCE: CAR4POA,CAR4POB,CAR4POC;
;**** APPLY SEALANT AND POSITION WINDSHIELD
CALL_SEQUENCE: CAR4POD,CAR4POE;
CAR4POD  [ 49]:> GOTO_TPOINT: CAR,TP16,NOP;

CAR4POE  [ 18]:> GOTO_TPOINT: CAR,TP2,NOP;
```

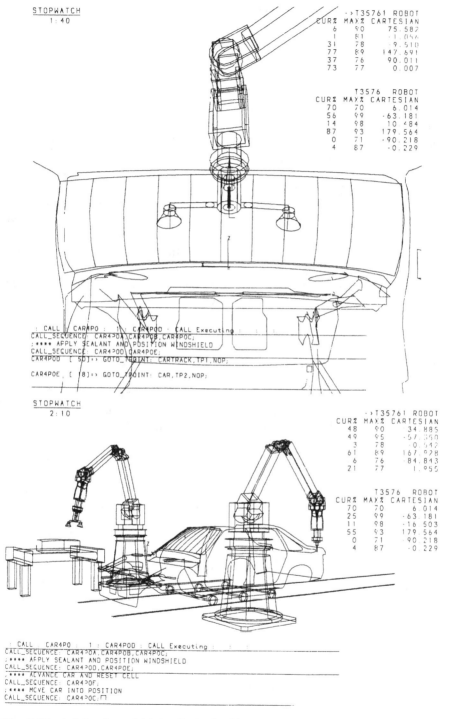

STOPWATCH
1:40

```
                            ->T35761 ROBOT
                   CUR%  MAX%  CARTESIAN
                     6    90     75.582
                     1    81     -1.052
                    31    78     -9.510
                    77    89    147.691
                    37    76     90.011
                    73    77      0.007

                             T3576  ROBOT
                   CUR%  MAX%  CARTESIAN
                    70    70      6.014
                    56    99    -63.181
                    14    98     10.484
                    87    93    179.564
                     0    71    -90.218
                     4    87     -0.229
```

```
: CALL CAR4PO : 1 : CAR4POD : CALL Executing :
CALL_SEQUENCE: CAR4POA,CAR4POB,CAR4POC;
;**** APPLY SEALANT AND POSITION WINDSHIELD
CALL_SEQUENCE: CAR4POE;
CAR4POD [ 50]=> GOTO_POINT: CARTRACK,TP1,NOP;

CAR4POE [ 18]=> GOTO_POINT: CAR,TP2,NOP;
```

STOPWATCH
2:10

```
                            ->T35761 ROBOT
                   CUR%  MAX%  CARTESIAN
                    48    90     34.885
                    49    95    -57.550
                     3    78     -0.542
                    61    89    167.928
                     6    76     84.843
                    21    77      1.955

                             T3576  ROBOT
                   CUR%  MAX%  CARTESIAN
                    70    70      6.014
                    25    99    -63.181
                    11    98    -16.503
                    55    93    179.564
                     0    71    -90.218
                     4    87     -0.229
```

```
: CALL CAR4PO : 1 : CAR4POD : CALL Executing :
CALL_SEQUENCE: CAR4POA,CAR4POB,CAR4POC;
;**** APPLY SEALANT AND POSITION WINDSHIELD
CALL_SEQUENCE: CAR4POD,CAR4POE;
;**** ADVANCE CAR AND RESET CELL
CALL_SEQUENCE: CAR4POF;
;**** MOVE CAR INTO POSITION
CALL_SEQUENCE: CAR4POC;
```

Fig. 7.12 A 'place' model joint limit checking
Source: McDonnell Douglas

simulation, the robot is moved around the cell to give an estimation of the cell cycle time and allow collision checking between the items within the cell. The simulation will also ensure that the robot joints are within their limits for the particular cell layout (see the output in Fig. 7.12 for a windscreen insertion installation).

There are a number of limitations associated with such models. The true robot dynamic behaviour is not modelled, so that there may be significant errors in the cycle time estimation and robot position. Automated collision checking is also not always possible, and may have to be done visually, especially in wireframe models.

Such models do, however, allow the development of better cell layouts and process cycle times, and can ensure that the correct robot is picked for the task.

In the simulation it is necessary to control the position of the robot, and this is done by defining the desired position and orientation of the tool. The simulator can then compute all the joint attitudes using data about the robot given when the model was created.

Having developed a program to drive the position of the robot model in the simulation, a natural development is to provide a post-processor that will produce a usable robot program which will control the position of the robot. These take the Cartesian world of the cell model, and using inverse kinematics (as discussed in section 4.6) obtain motion instructions for the robot joints. This data is then post-processed into motion commands in the target language. This is analogous to NC part programming from a CAD model of the object to be manufactured. Such packages are now available on a commercial basis. They are not, however, able to readily handle the conditional actions due to sensor inputs. The adoption of such model-based programming methods will be speeded by the introduction of more standardized CAD to robot controller to robot servo data transfer. IRData (a German initiative, the principles of which are shown in Fig. 7.13) attempts to do this.

It is necessary to adjust this 'theoretical' robot program produced in the ideal model CAD world for errors in the real workspace and in the real robot before the program can be executed safely. This is carried out by an interactive calibration program. Table 7.2 shows the modules of a commercial modeller. The usefulness of such systems is indicated by the problems associated with the reprogramming effort required in the reconfiguration of an automotive spot welding line – initial offline part program generation for a robot line of more than 100 robots gives considerable savings in line down time.

7.11 The Petri net

We now consider how we might model the logical interactions of the machines in our cell. The Petri net graph is being used by many as such

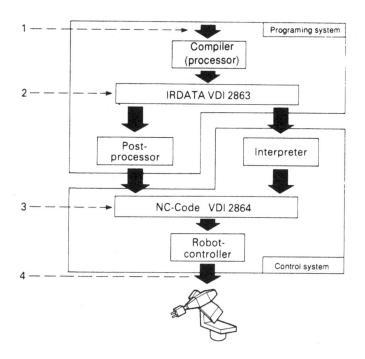

Fig. 7.13 IRData
Source: D'Souza *et al*

a model to represent the interactions between machines. This representation allows the designer to work through and understand the logic of his system for design, analysis and programming, especially when the system consists of a number of parallel activities. It is also being used as an input for automated control program generation.

Table 7.2 The modules of the MRS cell modelling and offline programming system
Source: McDonnell Douglas

Module	Task
BUILD	Set up models of robots and cell elements from a library of models
PLACE	Simulate the cell activity and check layout and robot capability
COMMAND	Generate robot program in target language
ADJUST	Calibrate cell model to real world, adjust the robot program

A Petri net consists of four parts; a set of places, P, a set of transitions, T, an input function, I, and an output function O. The input and output functions relate transitions and places. A Petri net structure, C, can be shown as

$$C = (P,T,I,O),$$

where $P = (p_1, p_2, \ldots, p_n)$ is a finite set of places, $n \geqslant 1$, $T = (t_1, t_2, \ldots, t_m)$ is a finite set of transitions, $m \geqslant 1$, $P \cup T = 0$, the set of places and the set of transitions are disjoint, $I: T \rightarrow P\infty$ is the input function, a mapping from transitions to places and $O: T \rightarrow P\infty$ is the output function, a mapping from transitions to places.

This theoretical framework can be represented as a directed graph. In the graph, each circle (place) on the graph represents a device, each vertical or horizontal bar represents a transition between states and a directed arc connects the two. A token (black disc within a place) moves around the system as events occur, transitions fire when all their input places are filled to indicate the output of the particular transition.

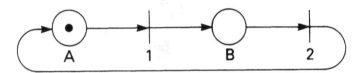

Fig. 7.14 A Petri net with two places and two transitions

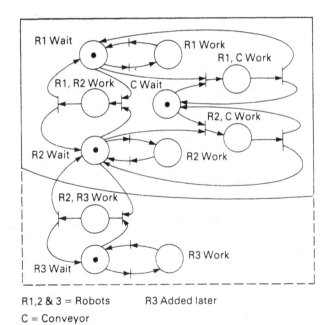

R1,2 & 3 = Robots   R3 Added later
C = Conveyor

Fig. 7.15 A Petri net of a cell
Source: Kakuza *et al*

The simplest net is shown in Fig. 7.14, and a more complex net representing co-operating machines in a cell in Fig. 7.15. Such nets can be extended so that for example the timing of each transition is modelled.

7.12 Cell justification

The usual test of the design of a manufacturing cell is that it should achieve a target cycle time for a particular set of part variants and can be financially justifiable. It is not the purpose of this book to indicate how to justify installations; the reader is asked to refer to this elsewhere. It should be recognized that all those who design and specify automation operate within the context of a company, and that each company has its own internal accounting practices which have to be used to justify the automation.

In this chapter we have indicated the shape of manufacturing cells as they are constructed at present. Future cells will show the increased use of complex sensory information (for example, much of the work applying machine vision to robotics has been directed to MIG welding) and increased parallel operation of the cell elements.

8   Assembly

After reading this chapter the reader should understand:

- the mechanics of assembly;
- the importance of the SCARA robot for assembly;
- some design for assembly and disassembly rules;
- the roles of dedicated and programmable assembly automation.

8.1 Introduction

One of the most well known workers in the field of automatic assembly, Boothroyd (1982), said that the assembly process accounts for 50% of work-in-progress, labour and inspection costs. The assembly process is responsible for most of the added value to a product. Assembly is therefore particularly commercially significant and offers real opportunities to reduce cost and lead time. As it is the area to which the previously produced parts come together it also shows up all the errors in previous production processes.

To indicate the complexity of the assembly operation it is worth examining what a human operator does to assemble two objects. First the person will locate, using vision, the two objects, one at a time. Then he will move first one hand to grip one part and then the other hand to grip the other. Using vision once again (and his manual dexterity) the person will bring the two objects together. He will then begin to put them together, first monitoring their relative position with his eyes, correcting this relative position with his hands and then detecting when they contact by using force sensing in his hands. As the process proceeds the person will monitor the insertion, the fundamental operation of assembly, using force and tactile sensing in his finger tips. I hope this indicates the demands that we are placing on assembly machines to replicate this manual process.

This chapter examines the assembly process from three distinct viewpoints. The first is an analytical one, which discusses some of the variables associated with the insertion as an aid to understanding the automation of that process. The second viewpoint is that of the designer, who must design for assembly to reduce the complexity of the assembly and disassembly tasks. The final section of the chapter briefly examines some of the techniques of automatic assembly.

8.2 Parts mating theory

The assembly process can be represented basically by the peg-in-hole insertion, that is the placing of a cylindrical shaft into a cylindrical cavity (see Fig. 8.1). Problems arise due to errors in the relative position of the peg with respect to the hole, and these errors are translational and rotational.

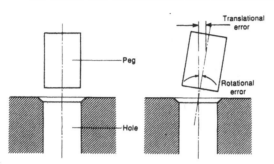

Fig. 8.1 The peg in hole problem

The insertion process has five essential stages which are shown in Fig. 8.2. These are (a) the approach, (b) crossing the chamfer at the edge of the hole, (c) making one point contact with the edge of the hole, (d) making two point contact with both sides of the hole and finally (e) sliding into the hole. The relationship between the angle of approach and penetration can be seen in Fig. 8.3 – the insertion life cycle.

(a)	(b)	(c)	(d)	(e)

Fig. 8.2 Five stages of assembly: (a) approach, (b) chamfer crossing, (c) one point contact, (d) two point contact, and (e) sliding in
Source: Whitney (1982)

To examine whether assembly will succeed or fail, it is necessary to consider the forces acting on the parts as well as the purely geometric considerations arising from tolerances for example. These have been analysed so that force measurements being made during the insertion process can, for example, be fed back to a manipulator so that the path of the manipulator can be controlled to allow the insertion process to proceed. Two phenomena are defined ('wedging' and 'jamming'), to describe situations in which the peg seems to stick in the hole during two-point contact and which prevent the insertion procedure from succeeding. Jamming will be considered first.

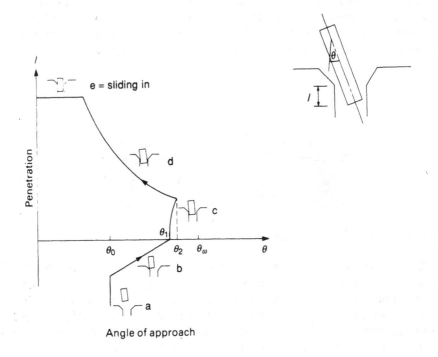

Fig. 8.3 Life cycle plot of insertion
Source: Whitney (1982)

8.3 Jamming*

Jamming is a condition in which the peg will not move because the forces and moments applied to the peg through the support are in the wrong proportions. A simple derivation of the problem is presented here.

We express the applied insertion force in terms of F_x, F_z, and M at or about the peg's tip, as shown in Fig. 8.4. We wish to find equilibrium sliding in conditions between the applied forces and the reactions f_1 and f_2

Making the first of many simplifying assumptions we ignore the angle of tilt of the peg with respect to the hole. Then, by resolving and taking moments about the second point of contact, the equilibrium equations which describe the peg sliding in are then given by

* This section is adapted from Whitney (1982); reprinted by permission.

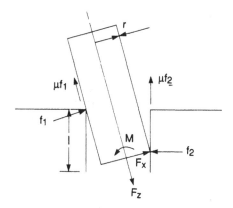

Fig. 8.4 Jamming two point contact
Source: Whitney (1982)

$$F_z = \mu(f_1 + f_2) \tag{8.1}$$

$$F_x = f_2 - f_1 \tag{8.2}$$

$$M = f_1 l - \mu r(f_2 - f_1). \tag{8.3}$$

Combining these equations yields

$$\frac{M}{rF_z} = \frac{l}{2r\mu} - \frac{F_x}{F_z}\left(\frac{l}{2r} + \mu\right). \tag{8.4}$$

λ is defined by

$$\lambda = \frac{l}{2r\mu}. \tag{8.5}$$

Then equation (8.4) can be expressed as a straight line

$$y = mx + c, \tag{8.6}$$

where

$$y = \frac{M}{rF_z} \tag{8.7}$$

$$x = \frac{F_x}{F_z} \tag{8.8}$$

$$m = -\mu(1 + \lambda) \tag{8.9}$$

$$c = \lambda, \tag{8.10}$$

which we can then draw on orthogonal axes.

If the peg is considered drawn the other way (the other case of the two point contact), the same equation results with

$$c = -\lambda. \tag{8.11}$$

To conclude the analysis we must consider the four possible one-point contacts, contact in the bore and contact at the corner with the peg leaning both ways. Two of these are sufficient to illustrate the analysis (Fig. 8.5).

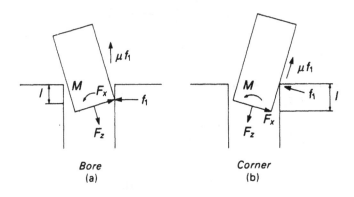

Bore
(a)

Corner
(b)

Fig. 8.5 Two (of four possible) one-point contacts
Source: Whitney (1982)

For Fig. 8.5(a), contact on the bore, and taking moments about f_1 and resolving horizontally and vertically, the equilibrium equations are

$$M + rF_z = 0 \tag{8.12}$$

$$F_x = f_1 \tag{8.13}$$

$$F_z - \mu f_1 = 0, \tag{8.14}$$

or, preparing to plot it on the same axes as those developed from equation (8.6),

$$\frac{F_x}{F_z} = \frac{1}{\mu} \tag{8.15}$$

$$\frac{M}{rF_z} = -\mu \frac{F_x}{F_z} = -1. \tag{8.16}$$

Similarly, for Fig. 8.5(b), contact on the corner, we have

$$M + lF_x + \mu rF_x = 0 \tag{8.17}$$

$$F_z - \mu F_x = 0, \tag{8.18}$$

or

$$\frac{F_x}{F_z} = \frac{1}{\mu} \tag{8.19}$$

$$\frac{M}{rF_z} = -(2\lambda + 1) \tag{8.20}$$

For the other two one-point contacts we can obtain

$$\frac{F_x}{F_z} = \frac{1}{\mu} \tag{8.21}$$

$$\frac{M}{rF_z} = 1 \text{ or } (2\lambda + 1). \tag{8.22}$$

It can be shown that these four points lie on the two lines which obey equation (8.6), restated as

$$\frac{M}{rF_z} = \pm \lambda - \frac{F_x}{F_z} \mu(1 + \lambda), \tag{8.23}$$

and are the endpoints of these lines because the lines represent the peg just sliding in, i.e.

$$F_z \geqslant \mu F_x \tag{8.24}$$

or

$$\frac{F_x}{F_z} \leqslant \frac{1}{\mu} \tag{8.25}$$

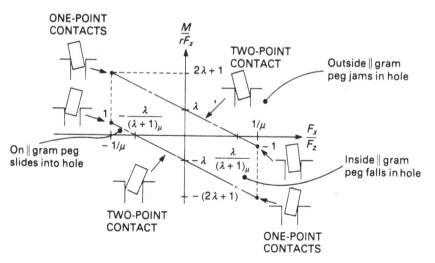

Fig. 8.6 The jamming diagram
Source: Whitney (1982)

for the right side one-point contacts, and similarly with a minus sign for the left-hand side ones. Larger F_x/F_z results in one-point contact jams.

All of the above can be summarized as Fig. 8.6 which plots the results in the forms of equation (8.6). The vertical dotted lines in the diagram describe a line contact. Figure 8.6 may be interpreted as follows: combinations of F_x/F_z and M falling on the parallelogram's edges describe equilibrium sliding in. Outside the parallelogram lie combinations which jam the peg, either in one- or two-point contact. Inside, the peg is in disequilibrium sliding or falling in.

8.4 Wedging

Wedging is also a condition in which the peg appears stuck in the hole, but unlike jamming, the cause is geometric rather than ill-proportioned applied forces. Wedges can be so severe that no reproportioning of the applied forces can cause assembly to proceed except by damaging the parts at their contact points. Wedging rarely occurs in automatically assisted assembly, where larger forces are applied because the larger forces push the peg through the jamming region by deforming the parts at the contact points.

To model wedging we must assume that at least one of the parts is elastic, although it is still stiff compared to any stiffness of the insertion device. In wedging, the two contact forces f_1 and f_2 can point directly toward each other, storing energy in the elastic part. This is possible if two-point contact occurs when l_2 is small, allowing the two friction cones at the contact points to intersect. Figure 8.7 shows one possible situation, in which l_2 is as large as possible and still allows wedging. The right side contact force does not point along an extreme of the friction cone,

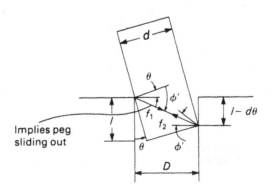

Fig. 8.7 A wedging condition
Source: Whitney (1982)

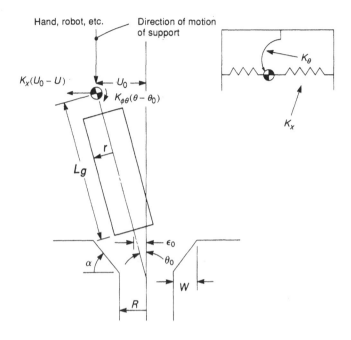

Fig. 8.8 Definition of terms for geometric parts mating
Source: Whitney (1982)

indicating that relative motion between the parts on the right side has stopped. The left side contact force points along the lower extreme of the friction cone, indicating that the left side of the peg is attempting to move out of the hole. (Recall that the direction of the friction force is opposite to the direction of motion.) This could occur if the peg has been pushed counterclockwise, elastically deformed at the contact points, and released.

To avoid wedging the peg shown in Fig. 8.8, with a compliant (sprung) support, the following result, indicating the magnitudes of the allowable rotational and displacement errors, must be obeyed.

$$\theta_0 + s\varepsilon_0 < \pm \, C/\mu \qquad (8.26)$$

where

$$s = \frac{L_g}{L_g^2 + K_\theta/K_x} \qquad (8.27)$$

and the non-dimensional clearance is given by

$$C = \frac{R - r}{R}. \qquad (8.28)$$

This result is shown in Fig. 8.9, together with the observation that for

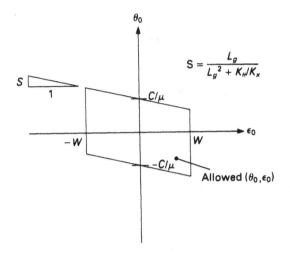

$$S = \frac{L_g}{L_g^2 + K_\theta/K_x}$$

Fig. 8.9 Geometry constraints on lateral and angular error to cross chamfer and avoid wedging
Source: Whitney (1982)

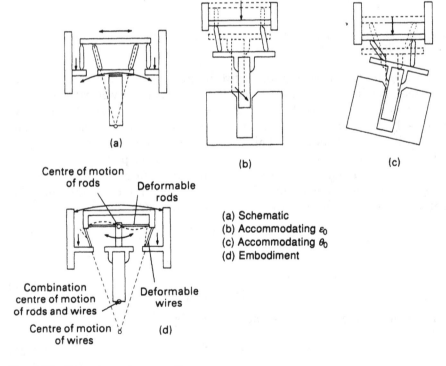

(a) Schematic
(b) Accommodating ε_0
(c) Accommodating θ_0
(d) Embodiment

Fig. 8.10 Remote centre compliance
Source: Whitney *et al.*

insertion to proceed the displacement error must be less than the width of the chamfer *W*.

8.5 Remote centre compliance

This analysis, because it has allowed the understanding of the problem, has led to the development of remote centre compliance (Fig. 8.10), a device that passively accommodates errors of displacement and rotation in assembly equipment, and thereby reduces the need for the use of sensors in, for example, robotic assembly. Remote centre compliance removes the compliance centre (the crossed circle in the diagrams) to the tip of the peg (L_g is very small).

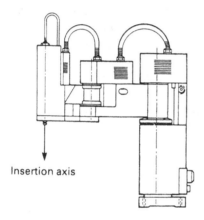

Insertion axis

ELEVATION

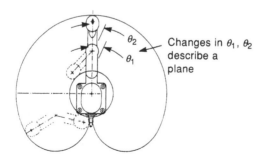

θ_2

θ_1

Changes in θ_1, θ_2 describe a plane

PLAN AND WORKSPACE

Fig. 8.11 A SCARA robot
Source: IBM

8.6 SCARA robots

Another device that has been specially designed to overcome such rotational and translational errors is the SCARA robot. The Selective Compliance Arm for Robotic Assembly (an example is shown in Fig. 8.11) is very stiff to rotations about its insertion axis, preventing the rotational errors from arising, and much less stiff to displacements in a horizontal plane allowing it to accommodate translational errors easily.

The machine has been specifically designed for two-dimensional electromechanical assembly, which consists essentially of motions in a horizontal plane to locate the component in space and insertions along the vertical axis. Figure 8.12 shows the ADEPT, a direct drive robot being used for 'odd-form' electronic assembly. The use of direct drives, the drive being an integral part of the robot joint rather than a motor and gear box, increases the speed and accuracy of the machine significantly. Machine vision is used in this particular installation to increase robot accuracy on a particularly demanding placement task.

In this section we have tried to identify some of the difficulties associated with assembly, and some of the mechanical designs that can be applied to assembly machines to reduce these difficulties. We now consider how the assembly process can be made easier by product design.

8.7 Design for assembly

Before discussing how we automate or organize the assembly process we must examine the use of design for assembly rules. These can be applied to significantly reduce assembly complexity and cost and are summarized below.

8.8 Rules for product design

These are the rules that can be applied to the complete assembly.

1. Minimize the number of parts. By reducing the number of parts in an assembly, the number of assembly tasks to be carried out is immediately reduced. This reduction also significantly reduces the organizational and transport problems associated with the assembly process.
2. Use standard parts where possible. The use of standard parts, such as the same fastener throughout an assembly, reduces the amount of tooling required and contributes to effectively reduce the number of parts in an assembly from an organizational point of view. The use of

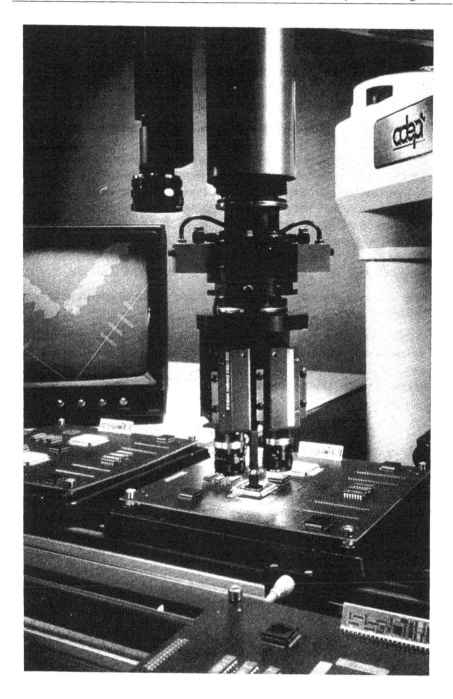

Fig. 8.12 Electronic assembly with a SCARA
Source: Meta Machines

standard, familiar parts also reduces the complexity of the assembly task by reducing the number of tools required.

The building up of an assembly can be assisted by the next four rules which help to locate the parts of the assembly in a stable way and use gravity to assist the assembly task.

3. Ensure that the product has a suitable base part on which to build the assembly.
4. Ensure that the base part has features that will enable it to be readily located in a stable position in the horizontal plane.
5. If possible, design the product so that it can be built up in layer fashion, each part being assembled from above and positively located (so that there is no tendency for it to move under the action of horizontal forces).
6. Try to facilitate assembly by providing chamfers or tapers which will help to guide and position the parts in the correct position.

To help minimize the complexity and expense of the assembly task apply the following.

7. Avoid expensive and time-consuming fastening operations, such as screwing and soldering. Use snap fits or other push together mechanisms where possible.
8. Make modular sub-assemblies.

8.9 Rules for the design of parts

These rules which can be applied to the individual parts of the assembly will assist reducing the parts picking, orientation and recognition problems in manual assembly, and parts feeding problems in automatic assembly.

1. Avoid projections, holes, or slots that will cause tangling with identical parts when placed in bulk in a feeder or bin. This may be achieved by arranging that the holes or slots are smaller than the projections.
2. Attempt to make the parts symmetrical, to avoid the need for extra orienting devices and the corresponding loss in feeder efficiency. (Ensure that parts cannot be assembled ambiguously.)
3. If symmetry cannot be achieved, exaggerate asymmetrical features to facilitate orienting or, alternatively, provide corresponding asymmetrical features that can be used to orient the parts.

The rules above are actually a combination of rules for design for assembly and for design for automated assembly. It should be noted that it is of value to apply these rules to all cases of assembly, as the design rule that assists automatic assembly is likely to assist the human operator.

If the reader actually has a real job to design for assembly or design for assembly automation, he is recommended to read and apply the work of Boothroyd *et al.* (1982) and Redford and Lo (1986).

8.10 Design for disassembly

There are increasing ecological pressures, such as the amount of landfill space available, that are leading to environmental legislation that will force suppliers to take back products at the end of their life. This legislation is forcing industry, particularly the builders of consumer electronics, to consider design for reuse, design for recycling and design for disassembly. Primary rules include:

1. Use environmentally friendly materials, such as thermoplastics that can be reused rather than thermosetting materials which cannot.
2. Reduce the use of heavy metals.
3. Do not use mixtures of materials and clearly mark those that you do use.
4. Use fastening systems that can be disassembled.
5. Design components that have a longer life than that of the product.
6. Define a logistics process that allows the effective collection of the old product.

Perhaps the most advanced proponents of this approach are photocopier manufacturers who already 're-manufacture' 25% of components for inclusion in new machines.

8.11 Manual assembly

As we have seen, assembly is a particularly complex task and can demand the use of force and tactile sensing, vision sensing as well as dexterity. Without doubt, the best device to carry out a variety of assembly processes is a human operator at a bench or in the assembly line because of his intelligence and sensory abilities. Human operators are very difficult and expensive to duplicate with automation and automation is only financially justifiable in large production quantities. After careful design for assembly it may be preferable not to have assembly automation at all and use solely manual assembly.

8.12 Dedicated assembly automation

A typical machine for mechanical component assembly has a number of separate workstations set out in line or around a curve or circle (see

Fig. 8.13 for a rotary table based installation). Such machines are essentially lines with serial transfer. The assembly itself is built up from the base component as it moves from one work-station to another. Transfer within the machine is automatic and component orientation is maintained at all times. The particular installation illustrated includes a spring making machine. Springs are particularly difficult to handle because of their elastic properties and tendency to tangle. It is now becoming usual to manufacture springs at the assembly machine so they are not released until they are actually placed into the assembly.

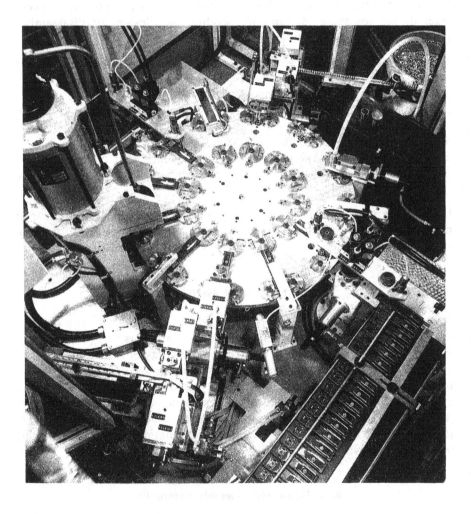

Fig. 8.13 A rotary indexing machine
Source: John Brown Automation

Component orientation for all forms of assembly installation is obtained using bowl or vibratory feeders (Fig. 8.14), or by purchasing ordered components in, for example, magazines or bandoliers. The bowl feeder works in the following manner. The vibrating bowl is loaded with disordered components, a helical track around the outside of the bowl carries the components from the bottom of the bowl to dedicated tooling at the top of the track, which will only permit components in certain orientations to escape from the bowl. Components in other orientations fall back into

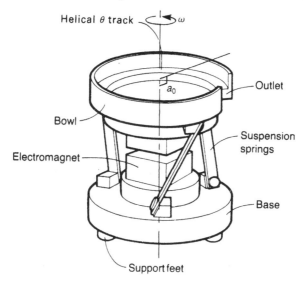

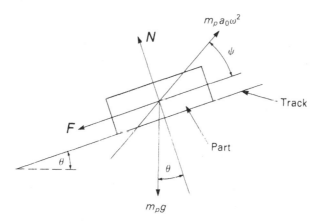

Fig. 8.14 A bowl feeder
Source: Boothroyd (1981)

the base of the bowl. Using the nomenclature in Fig. 8.14, the conditions that cause the components to rise up the track by 'hopping' can be determined by resolving along the track, recognizing that the acceleration provided by the motion of the bowl will overcome friction:

$$m_p a_0 \omega^2 \cos\psi > m_p g \sin\theta + F \tag{8.29}$$

and normal to the track and multiplying by the friction coefficient μ_s, to give

$$F = \mu_s N = \mu_s (m_p g \cos\theta - m_p a_0 \omega^2 \sin\theta), \tag{8.30}$$

giving the condition for forward motion as

$$\frac{a_0 \omega^2}{g} > \frac{\mu_s \cos\theta + \sin\theta}{\cos\psi + \mu_s \sin\psi}. \tag{8.31}$$

The key to the automation of assembly is the feeding of sufficient parts to the assembly workheads.

The assembly operation itself is automated – usually as modular workstations such as a cam driven pick and place device, nut runners or a machine screwdriver. Such modules are general designs that can be configured into a range of machines. Figure 8.15 shows an in-line linear transfer machine configured from modules.

8.13 Programmable assembly automation

Programmable assembly automation can be either based around robots or special purpose programmable assembly machines.

Most programmable assembly installations using robots are cells using a single machine or lines. Robot installations for mechanical assembly are usually Cartesian or SCARA robots with a linear transfer between them. Mechanical programmable assembly automation is rarely flexible and installations are frequently dedicated to a single product or small product range.

Programmable assembly is most frequently encountered in the assembly of electronic components to circuit boards, where programmable component insertion machines and Cartesian and SCARA robots are fairly widely applied. The two-dimensional electronic assembly processes are much easier to automate than the three-dimensional mechanical ones. It should be recognized that electronic assembly is essentially a mechanical process and of such a small scale that an automated process is frequently the only practicable option.

The programmable insertion or placement machine is a dedicated programmable device specially configured to place most electronic components. They are typically large Cartesian tables. An example of such a machine for electronic surface mount devices (SMDs) is shown in Fig. 8.16. Such machines are usually fast and it is usual to only use a robot, which is comparatively slow, for 'odd-form' components that cannot be easily placed with another machine.

Fig. 8.15 Dedicated assembly transfer machine
Source: John Brown Automation

Fig. 8.16 Electronic component placement machine
Source: Dynapert Precima

Fig. 8.17 A separated motion machine

Recent designs of flexible assembly machines are based around the concept of a 'separated motion machine' as illustrated in Fig. 8.17. In a separated motion machine single machine axes are suspended from an x-y plane. Each axis is dedicated to a single task, for example, placing, screwing or soldering, and is allowed to move around the x-y plane, the active end of the axis being held above a second x-y plane on which the workpiece is situated. Motion of the workpiece, component feeders and the axes above the machine allows the assembly process to proceed. Such systems frequently include vision systems to improve their accuracy and are comparatively easy to reconfigure to assemble different products. They are particularly intended to be used for small batch electro-mechanical assembly of, for example, aerospace electronic equipment.

9   The integrated factory and systems

After reading this chapter the reader should understand:

- the characteristics of the automated factory;
- flexibility;
- manufacturing as a hierarchy;
- the variants of the automated factory;
- the use of discrete event simulation for system design.

9.1 Introduction

In this chapter we will examine how the discrete devices and cells that we have described so far in the book can be put together into larger manufacturing systems.

Recall that our essential aim is to design and operate a profitable manufacturing facility which produces an increased variety of components and rapidly, right first time, and with minimal manning. This need for increased variety, small batches, short lead times and fast changeovers has led to the use of programmable automation rather than more conventionally automated facilities. This business need can be summarized as the need to manufacture our products better, faster and cheaper!

It is worth noting that our system is not unmanned, and it has the minimum manpower associated with efficient and economic manufacturing system performance.

The system that we produce must be integrated. Manufacturing systems integration can be defined as the process of ensuring that all the elements of the system – manufacturing equipment, computers and people – work together to achieve the system purpose. Well integrated systems have smoothly functioning interfaces so that, for example, information, parts and tools pass from one sub-system to another efficiently to give a responsive system. One of the implications of this is that the facility and the product should be developed together. This parallel development is sometimes called 'simultaneous or concurrent engineering' and uses many of the simulation tools described elsewhere in the book.

Many of the systems that we will examine in this chapter are based on metal cutting machine tools and are in either high volume areas such as the automotive industry or high added value areas such as the aerospace industry. The metal cutting industry is many years ahead of other discrete parts manufacturing industries largely because of the sophistication and programmability of metal cutting machine tools. It is likely that other industries will learn from the systems that have been used for metal cutting in their search for appropriate systems to manufacture their own parts.

9.2 Flexibility and reconfigurability

Before we begin to look at some of the systems that have been installed we must examine the concepts of flexibility and reconfigurability.

9.2.1 FLEXIBILITY

Flexibility (sometimes short term flexibility) refers to the ability of a manufacturing system to process a number of different parts from a pre-defined group of parts. For example, a system that makes a hundred different parts is more flexible than a system that makes twenty different parts. Manufacturing systems, even the most programmable, are usually designed to produce a pre-determined group of parts.

Within this definition of flexibility, further refinements are possible; these include 1. mix flexibility, the ability of the system to cope with a variety of components or materials, 2. routing flexibility, where the system is complex enough to allow the route between machines to be changed, and 3. volume flexibility, where the system can efficiently cope with changes in required output.

9.2.2 RECONFIGURABILITY

Reconfigurability (sometimes long term flexibility) can be taken to refer to the ability of a manufacturing system to process a group of parts other than those for which it was designed, or which involve a significant product changeover. This measures how difficult is it to re-tool and re-program the system to make a different set of parts. This is particularly significant if the manufacturing facility is required to have a life longer than that of the product. Present short product life cycles emphasize the importance of this.

Absolute definitions of these concepts are difficult to make but they provide a subjective method of comparing rival system designs. It should also be observed that the manufacturing facility is required to produce the product variety that the customer demands. It is usual to add this variety as late as possible in the manufacturing process.

9.3 A hierarchical view of manufacturing

The manufacturing task within a company can be viewed as a hierarchy of a number of autonomous operating levels, and the technologies and organizations that correspond to these. Figure 9.1 shows an example of

1 — Source-secure.   — Manufacturing company
Commercial unit of company (stocks quoted)
Product: range of cars and trucks
2 — Logistic   — Manufacturing plant
Planning unit of company
Product: range of power trains
3 — Autonomous   — Product manufacturing system
Product: power train
(engine, clutch, gearbox and differential)
4 — Market sensitive   — Flexible production system, sub-contractors
Product: variety of complete engines with bought out sub-assemblies
(engine plus carburettors, fan belt, pulleys, alternator, distributor, etc.)
5 — Manufacturing process   — Sub-assembly manufacturing system
Product: basic engines
(block, head, crankshaft, camshaft, conrods, pistons, valves, etc.)
6 — Operation   — Component (GT group) manufacturing system
Product: basic critical components
e.g. camshaft
7 — Primitive   — Machine
                            |
                        Process
Product: part of geometry of component
e.g. cam geometry.

Fig. 9.1 Levels of operating autonomy and the product produced at each level
Source: Ruff (1985)

these levels in an automotive company and the sorts of product associated with them. Figure 9.2 shows the technologies that could be used to produce products at these levels. Such a hierarchical view demonstrates how the company decisions can be decentralized to the business centres of the company. This also indicates that the manufacturing technology decisions can be synthesized and integrated from descriptions of the processes required to make the parts that the business has decided that the facility will produce.

Such a hierarchical view means that the control of the facility can be hierarchical and that plant and machines can be installed and proved as individual devices. They then can be progressively integrated, from the 'bottom-up' into a complete and complex manufacturing system, this is the 'functional integration' approach.

In this book we are essentially concerned with levels three to seven of Fig. 9.2. Practical programmable manufacturing systems are being installed as such from level seven to level five. This chapter will examine some of these installations. The rest of the hierarchy is at present usually implemented in less automated manner. Fortunately, this allows the company to react dynamically to the changes in the competitive environment.

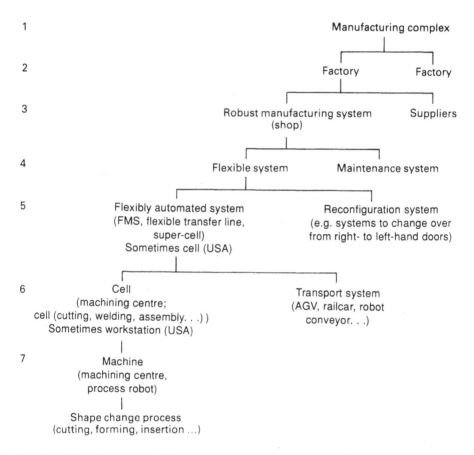

Fig. 9.2 Manufacturing facility and technology hierarchy
Source: Ruff (1985)

9.4 Generic elements of manufacturing systems

A manufacturing system can be considered to have three basic elements together with the people within the system:

1. the machine tools and manufacturing processes themselves;
2. the transport and handling (including storage) sub-system for parts and tools;
3. the control sub-system which includes the controlling elements, sensors and people.

This chapter will indicate some of the processes that have yielded so far to programmable factory automation approaches and some of the transport sub-systems that have been used in particular applications. The final chapters will concentrate on the control aspects. We must remember that the selection of the groups of parts for many of the systems and the concepts that have led to the systems design are actually based on the group technology approaches described in Chapter 2.

9.5 Direct numerical control

One of the precursors of current manufacturing systems approaches is direct numerical control (DNC).

In the early days of NC, the machine control program was usually prepared as a magnetic or, more usually, paper tape on a computer separate from the machine. This tape was then carried to the shop floor and loaded into the machine using a tape reader resident in the machine. In DNC installations the individual machines are linked to a central computer which stores all part programs on a disc. When an operator wishes to run a part program, he can call it at the individual machine and it is downloaded from the main computer to the particular machine. DNC installations are usually encountered where there are a large number of NC machine tools running a large number of part programs. The aerospace industry contains a number of examples. A schematic DNC system is shown in Fig. 9.3.

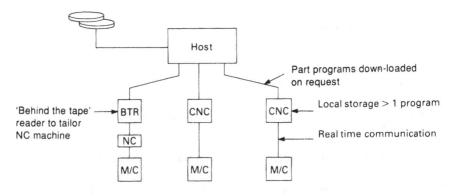

Fig. 9.3 A DNC system

9.6 Flexible cells

We have already seen in Chapter 7 that cells can be one of the starting points for a system as cells are a cheaper and perhaps more versatile solution than a large system. Large systems are mainly applied in the automotive and aerospace industries, or the Japanese machine tool industry, though small systems and cells have been applied in many industries.

9.7 Flexible transfer lines

Flexible transfer lines, sometimes flexible flow lines, generally take one of two forms – either robots carrying out a process linked by a conveyor or pallet shuttle, or machining centres linked in a similar way. The components in a system always move along the same path. A flexible transfer line usually has limited product variety and is often encountered in automotive applications. They can be reconfigured for another family of components by reprogramming the individual machines and by modifications to the fixturing, allowing the system to manufacture a new generation of products. Programmability allows product development and design changes during plant start-up, and this can dramatically reduce design to product lead times.

9.8 A robot transfer line

Perhaps the best example of a robot transfer line is the unpainted automotive body shell (body-in-white) spot welding line. This is the largest application of robots to date in the manufacturing industry. A typical installation can be represented by the similar Ford Sierra Lines installed at Dagenham in England and Genk in Belgium (Fig. 9.4). The individual elements of the body are prepared on lines which link together and then allow the production of the complete body assembly. The lines are capable of manufacturing both the Sierra and Fiesta and there can be up to 16 variants of each – note also that the line was commissioned for the Cortina.

The programmability of the robot allows the same machine to be used to place spot welds in the wide variety of positions required by each particular model in the wide model range. Also the increase in quality associated with the reproducible welding parameters and consistent weld position allows a decrease in the number of welds placed, as the product designer has more confidence in the manufacturing process.

Transfer lines using robots for mechanical assembly are usually difficult to reconfigure for a different range of products because of the requirements for extensive development of parts and gripper tooling.

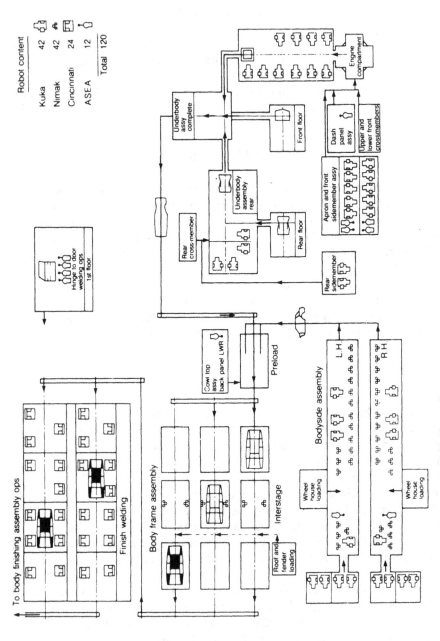

Fig. 9.4 A robot transfer line – the Ford Sierra line

Fig. 9.5 A CNC machine transfer line
Source: Cross International

9.9 A CNC machine tool transfer line

The first flexible system ever built, the Molins System 24, was described by its builder, D. T. N. Williamson (1967), as a flexible transfer line in 1967. The system essentially consisted of seven specially designed double spindle NC milling machines linked by a pallet conveyor.

Systems based on CNC machines linked by conventional transfer devices are increasingly applied. These system are generally more dedicated to particular products than 'fully fledged' FMSs. The programmability of the CNC machine tools allow reconfigurability from product to product provided the tooling and fixtures allow this. Figure 9.5 shows such a system for the machining of automotive components using three axis NC machines.

In a flexible line the machining required on a component is usually split between the machines in the line, so that each machine has a short cycle time. This is in contrast with an FMS, where once a component is on a machine it is usual to complete all the machining required on that particular machine.

9.10 A SMD assembly transfer line

Figure 9.6 shows surface mount device placement machines arranged as a transfer line. Although the handling equipment in the line only

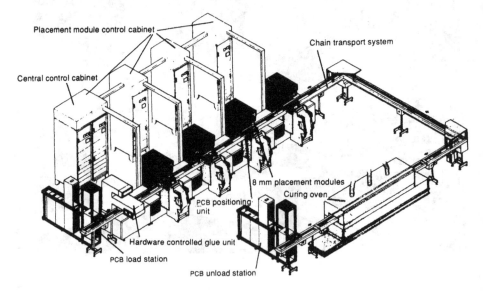

Fig. 9.6 A SMD placement machine transfer line
Source: Mullard

accommodates a small variety of circuit boards, the placement machines allow a very high variety of component positions and types. With the reducing scale of surface mount devices, it is usual for such machines to include vision systems to ensure accurate placement of the correct components.

9.11 Flexible manufacturing systems

The Japanese FMS builder T. Yamazaki gives the following definition of a flexible manufacturing system: 'It consists of three or more machining centres, turning machines, fabricating centres, or the like, equipped with flexible automatic loading/unloading/transfer devices and a method of monitoring tool conditions and replacement. The entire production scheduling and machining process is automatically supervised by a computer system.'

There are a variety of architectures used in the construction of FMSs and these are discussed below.

9.12 A monolithic flexible manufacturing system

A monolithic FMS, as exemplified by Makino at Atsugi, consists of programmable machines, a single parts transport sub-system, a tool transport system and storage for tools and components, all under the control of a hierarchical computer system. Figure 9.7 shows the Atsugi system, which consists of 10 machining centres, a wash station, a manual pallet loading area, a complex automated guided vehicle sub-system for transport, a warehouse, and a tool identification and handling system.

'Monolithic' implies that the whole system was conceived, built and financially justified at one go. The size and complexity of such systems makes them difficult to design, install, commission and justify financially. These systems are suited however to high product variety.

Such systems are sometimes called random access systems, as they allow access by any component to any machine at any time, and because of this they can have rescheduling flexibility.

The system makes parts for machine tools. 200 parts requiring 320 operations can be made by the system. The system operates on a batch of one principle (a batch of one is a pallet) and has a warehouse capacity of 550 pallets. It is capable of unmanned operation for up to 72 h. Tool breakage is checked by acoustic emission for small diameter tools and probing. Broken tools can be changed automatically. If, however, a tool breakage occurs during the day shift a new tool is loaded manually, as this is much quicker. Four or five men are required to run the system

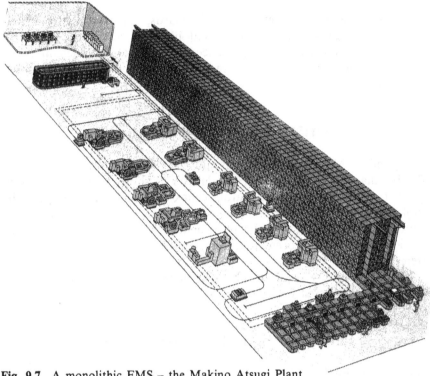

Fig. 9.7 A monolithic FMS – the Makino Atsugi Plant
Source: Makino Milling Machines

during the day – one man tool-setting; two men work-setting; and two to three for maintenance, programming and system monitoring.

Such systems have been constructed to produce parts for machine tools, automotive engine and transmission components, valves and prototype parts. These systems were often conceived before the mechanical and computational complexities of such systems were recognized.

9.13 Cell-based systems

Because of the complexity and difficulties associated with the one-off design, financial justification, installation and commissioning, and operation of large systems, FMSs of linked cells have been constructed (see Figs 7.6 and 9.8). These are sometimes called super-cells. They can be constructed cell by cell and linked with an inter-cell transport system at a later date. Cell-based approaches also allow the progressive evolution of the activity cell by cell. Fully automated cells and manually controlled cells are also more easily integrated.

Such systems are usually constructed for smaller parts families than monolithic FMSs, and usually have a number of cells within them carrying

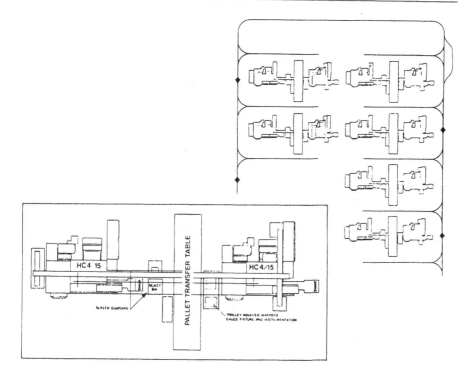

Fig. 9.8 A cellular turning FMS
Source: TI Machine Tools

out a similar function. They also do not operate on batch of one principles. There is redundancy within the system to cope with sub-system failures. The sub-systems can have a variety of functions such as cutting, welding or assembly.

The use of this architecture implies the use of at least two separate transport systems, one within each cell and one for the complete system. The cell transport system is usually a robot or railcar and the system transport system an AGV, so that typical systems are lathes tended by robots, linked by AGVs (refer to Fig. 7.6) or railcar-linked machining centres linked with other similar cells by AGVs.

Cellular systems have been constructed to build automotive components, turned parts, and sheet metal components. AGVs, carrying partially completed assemblies, have been used to link robot welding cells and manual assembly cells in automotive body-in-white manufacture and in engine assembly respectively.

It is apparent that cellular systems can be constructed gradually from their elements, and this is currently the preferred strategy for system design. The steps in the growth of a small railcar system are shown in Fig. 9.9, such railcar sub-systems can be linked with AGVs into even larger systems.

Stage 1 — Software options DNC CADCAM

The stand-alone machine, with its minimum manned capability allows the user to build-up experience in palletising work and presenting mixed batches of work to the machine. The FM 100 can have two or four pallets. The latter is a useful configuration if there is likely to be an extended time span between stage 1 and 2.

Benefits
A CNC machining centre with all the essential elements for extended periods of unmanned operation. Shop floor personnel can rapidly assimilate the principles of the new technologies.

Stage 2 — Software options DNC CADCAM

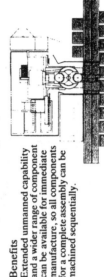
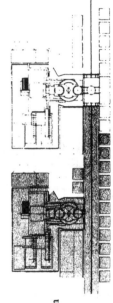

The addition of a rail-guided vehicle, and up to 15 pallet stands enhances performance in two ways. The range of components that can be handled — without operator involvement — increases. And, the periods of unmanned operation can be greatly extended.

Benefits
Extended unmanned capability and a wider range of component can be available for immediate manufacture, so all components for a complete assembly can be machined sequentially.

Stage 3 — Software options DNC CADCAM

Additional machines up to four, each with up to 15 pallet stands can be added. At this level, pallets are dedicated to individual machines. Load/unload is carried out at a load/unload station allocated to each machine. The rail guided vehicle responds to calls from the individual machines' control system.

Benefits
Increased manufacturing capacity from additional machines with the ability to further reduce inventory.

Stage 4 Software options DNC CADCAM

The same system hardware parameters as stage 3 plus an independent transport control. Any pallet can now be loaded to any machine. Dedicated load/unload stations improve operator efficiency. The transporter works to priorities established by the operator instead of answering machine calls for the next component on a queuing basis.

Benefits
Improved productivity from balanced scheduling leads to increased cell efficiency.

Intelligent car control, VDU and keyboard

Stage 5 Software options DNC CADCAM, host cell computer, link to business systems, CIM.

With the addition of a host computer the cell can be expanded to seven machines, inclusive of support operations including inspection and wash facilities. A host computer up-grades the system to exploit all the facets of Automated Manufacturing Technology.

Benefits
The final step to AMT. Linked into CAD, MRP and other business packages, the cell offers the user the fastest product development combined with rapid re-direction of resources to reflect changing market patterns.

Intelligent car control, VDU and keyboard

Fig. 9.9 A system growth path
Source: KTM

9.14 Virtual systems

The concept of the virtual flexible manufacturing system has been introduced by the system architects of the NIST for their AMRF (The National Institute of Standards and Technology Automated Manufacturing Research Facility).

The AMRF consists of a number of workstations (cells) of different function linked by a comprehensive, many path transport system (Fig. 9.10). By selecting an AGV route between particular workstations, 'any' type of manufacturing system (consisting of many processes in any order) can be synthesized, to allow the production of a very high variety of parts from the cellular architecture. This is known as a virtual system. This contrasts with the more dedicated cellular system designs presented above.

It is difficult to implement such a system of mixed technologies, due to problems in the development of the production process technology and the construction of a general transport system capable of carrying all the different shapes of product.

9.15 Hybrid systems

The phrase 'hybrid system' has been coined to describe a largely automated system that has a number of processes within it that are still carried out by an operator. Perhaps the most significant limitation of programmable factory automation is that some processes have not yet yielded to inexpensive automatic control. If we wish to include such processes it is necessary to use an operator with a craft based skill.

It may also not be financially justifiable to automate transport in a system. An example of this is the use of operators as a transport sub-system in a high variety prototyping facility, where the balance of the system is a conventional automated facility. The transport task would be very complex, and consequently difficult and expensive to automate flexibly.

An example of a hybrid system, with an unusual embodiment, has been constructed by Hitachi (Fig. 9.11), which uses a local area network to send commands to terminals. Messages on the terminals pace the production of manual operators by calling for the manufacture of components. The raw materials and finished components are transported by AGV.

Direct numerical control systems are another example of this philosophy.

9.16 People in systems

As system designers we have to think carefully about what we can and cannot automate, and where we can best use people, particularly where

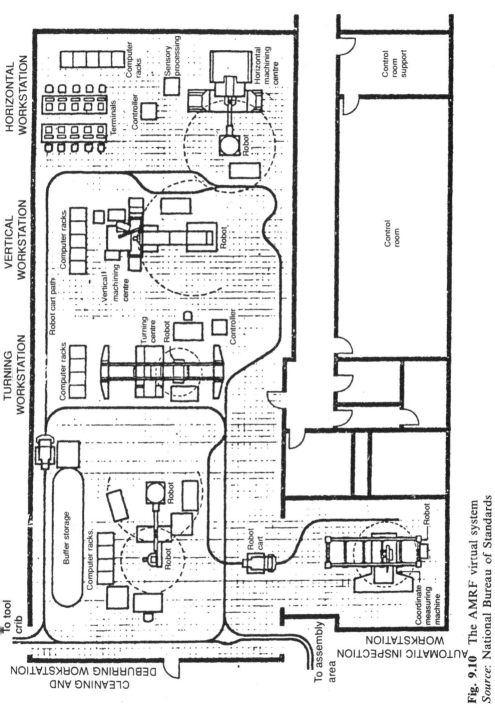

Fig. 9.10 The AMRF virtual system
Source: National Bureau of Standards

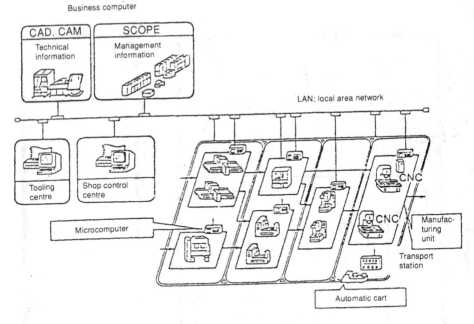

Fig. 9.11 A Hitachi hybrid system
Source: Shimotashiro *et al.* (1984)

they can play a role in integrating the system. As we have allowed ourselves to build only financially justifiable automation, we must still retain a skilled labour force on the shop floor for maintenance, system patrol, and for system recovery to ensure that the system does not go out of control.

We will, however, often require our system to be unmanned for some of its operating time. Because of the large investment demanded in factory automation we require the facility to be productive for as much time as possible. This implies that we will wish to run unmanned on night shifts, over the weekend and through breaks. This requires the automation or management of 'tacit tasks'.

The operators in a manufacturing facility carry out many, at first sight, trivial tasks that are critical to system viability. These are so-called tacit (unspoken) tasks. Many of these tasks would be very difficult and expensive to automate. Examples are:

1. *Orientation*. When an operator loads a part at the entry to the system he orientates it for the balance of the mechanical equipment. The automation of this using bin picking robots with vision and tactile sensing is currently impracticable.
2. *Inspection*. The operator will also usually not attempt to load a defective part into the system. An automatic facility might attempt to do this possibly damaging the system and producing scrap.

9.17 Simulation for system design

The manufacturing systems that we have described are very complex, as are the sub-systems within each system, and it is necessary to model the operational behaviour of such systems as part of the design process. The particular design problems that we have are unstructured and involve quantitative, qualitative and spatial data. Computer interactive, graphic discrete event simulation or modelling is a well established tool for this modelling task. Figure 9.12 shows the output from such a simulation.

Discrete event implies that the behaviour of the system is modelled in a step-by-step, event-by-event way, so that the system has transitions from

Fig. 9.12 Graphic output of a simulator
Source: Istel

state to state, rather than its behaviour varying as a continuous function. Computer simulation is experimentation on a computer-based model of a system. The model is used as a vehicle for experimentation usually in a trial and error way to demonstrate the likely effects of changing design variables.

A simulated experiment has many advantages over optimizing the design of a real system during the build stage. Real experiments consume the time of expensive skilled manpower and equipment, whereas simulation can be speeded up and, more importantly simulated experiments are safe, as they do not lead to irreversible damage to the real system.

The most common approach used in the UK is to use a three-phase simulator. The three phases are:

1. Phase A, a controller incrementing time. It is also sometimes known as the time advance or time scan. Its operation involves searching through the system description to find the time of the next event and then moving time forward to that point.
2. Phase B, that includes events that occur regularly in time. The simulator executes only those that are due as identified by the A phase. These are sometimes known as bound or book-keeping activities.
3. Phase C, conditional (control or co-operative) events. The simulator attempts all the C activities in turn, executing only those whose conditions are satisfied.

In this way the simulator moves the entities in the model through time, executing events and collecting the results of these events. Figure 9.13 shows a flow chart of a three-phase simulator operation.

An entity is an item that moves from one discrete state to another as events occur during the progression of time. These entities are the elements of the system that are being simulated. An example of an entity is a car body passing down a production line. Some entities are permanent (that is, they always remain within the system, such as a machine), and some are temporary and pass through and out of the system, such as the car body on the production line. Those entities that are acted upon are passive, e.g. car body, and those acting on others are active, for example, the machine. Entities can be grouped into classes of similar ones. Entities also have attributes which can be used to convey information about them, i.e. colour attributes sub-divide classes. Entities can also be placed into sets. Entities change state during the simulation, and in these states may be represented as sets – the current membership of sets of the system can represent the state of the system at a particular point in time. An example of this is the set of bodies waiting to enter an element of the production line – giving the state of that element of the line.

Now let us consider the input data necessary for model formulation. We must have a representation of all activities within the system. We must

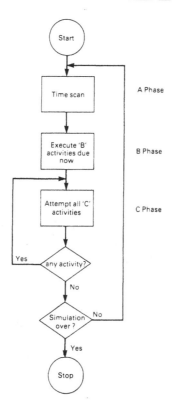

Fig. 9.13 Three-phase simulator flow chart
Source: Marr

include details of the processing devices, the types and numbers of machines, transport devices, buffers and jigs, fixtures and pallets.

There must be a time model of all the activities, for the process itself, for setting, for breakdown (mean time between failures and a likely distribution) of machines and tooling, and for recovery from this breakdown.

We must also specify the system start-up rules, the transport rules and the control rules governing machine interaction (for example, if an up-line machine has stopped, the downline machine may have to stop). It is important that the start-up conditions of the model are correctly specified as they significantly condition the way the model subsequently behaves.

The system overall control algorithms must be written. Inevitably, this starts at a naive level and it is developed as the simulation proceeds by intuitive interaction.

The simulation is then run with time accelerated and the model and design refined by designer interactions with the model – this is usually referred to as trying 'what-ifs?'. In this way the sorts of variable listed below can be evaluated:

- numbers of machines,
- numbers of robots,
- number of conveyors,
- numbers of AGVs,
- sizes of WIP and inventory buffers,
- system control algorithm development,
- queue length at system elements,
- system output.

Systems are increasingly becoming available that combine graphical features of the geometric simulation systems described in Chapter 7 with the logical features of the discrete event systems described here. The added value as a design tool provided by the geometric elements of the simulation is not clear when such simulation tools are applied to manufacturing systems designs that do *not* require the study of detailed machine kinematics and the determination of machine cycle times. They can however have a significant role in operator training and system marketing to senior management.

Critical issues in simulation are model validation and the time taken to develop the model. The designer must be sure that he is simulating exactly what he wishes to and that his simulation is not giving him erroneous results – all this is aided by the graphic output of the simulator. This is particularly difficult. Also the time taken to develop a model can be long, as a result of the effort and time taken to collect reliable data for the model.

9.18 Application areas

The opportunities for the less expensive production of reduced batch sizes using programmable automation (when compared with dedicated automation and manual methods) is indicated in Fig. 9.14. The actual application areas of some of the programmable systems presently applied in industry is indicated in Fig. 9.15. This shows the variety of parts (number of different parts) made in a system against the production volumes. This figure is not to be regarded as an absolute description of the application of such technologies, but is only intended to indicate current trends. It shows that most robot applications are usually for very high production volumes, and generally that turning systems and flexible transfer lines have higher production volumes than prismatic monolithic systems. The figure also indicates that the higher the throughput of a system the lower the variety.

It has often been said that, for a particularly complex manufacturing task, the disciplines that have to be placed on the balance of the

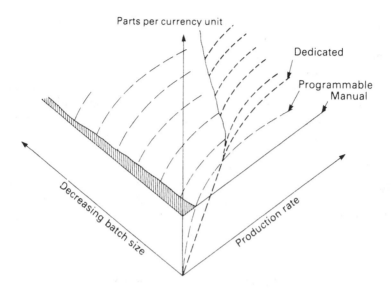

Fig. 9.14 Opportunities allowed by programmable automation
Source: Sabin

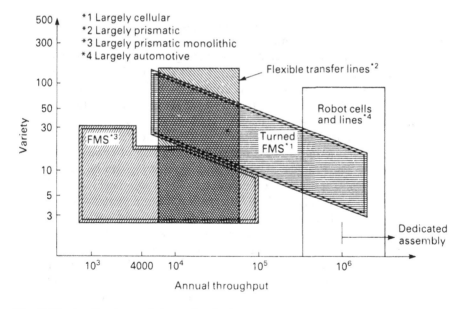

Fig. 9.15 Applications of the technologies

manufacturing facility (to allow the effective installation of the large programmable system) give as much of a productivity gain, with minimal investment, as the new system.

The installation of any complex system must be carefully considered, and should be carried out step by step. This will ensure that all is well with each individual step and will allow the profits from the earlier steps to fund the later ones.

Many of the projects described in this chapter have high technical and financial risks. Unsuccessful projects of this size, even within the automotive and aerospace sectors, can be permanently damaging to the trading position of companies. When considering projects to be implemented in your own company bear in mind '*caveat emptor*', let the buyer beware. Any project of this magnitude should be part of a complete manufacturing strategy for the whole company and should stand or fall on its own merits. If government grants have to be used to justify the system financially it is unlikely that it is the correct commercial and technical solution. Be pragmatic, and only use technology appropriate to the task.

9.19 The integrated enterprise

Since the preparation of the first edition of this book a number of significant culture changes have taken place within manufacturing in the West.

The first of these followed an 'after Japan' view of the world which promoted the use of JIT and continuous improvement philosophies; application of these techniques to Western factory systems reduced inventory, improved customer supplier relationships and quality (when coupled with the application of statistical process control techniques and total quality approaches). This led industry to turn away from plant-wide CIM solutions and concentrate on making its current generation of products well.

It was then recognized that Western industry needed to speed up its new product introduction process by the introduction of concurrent engineering practices. The practice of concurrent engineering can be significantly supported by the application of computational tools such as CAE and CADCAM; it also requires close integration of the manufacturing and design activities within an organization. This, combined with the need to exchange data with suppliers and to include marketing and sales explicitly, links with an understanding of manufacturing capacity, for example. These push us back towards the use of CIM tools within a large organization – the integrated enterprise. I think that we now understand better that we need to structure this enterprise into, literally, 'manageable' elements that have some degree of autonomy and are integrated where necessary by computer systems and people systems.

10   Computer control of manufacturing systems

After reading this chapter the reader should understand:

- the use of computers for system control;
- the fundamentals of hierarchical control;
- communications systems for manufacturing;
- the use and limitations of local area networks for manufacturing control;
- the practical embodiment of control systems.

10.1 Introduction

In the next two chapters we will look at the use of computers in the control of manufacturing systems and cells. This chapter will concentrate on hardware and the next on software.

The task of controlling automated batch manufacturing is the management of a complex collection of machines. This task is at present one of co-ordination of machines rather than control of the position of individual machines in space. The position of the machine in space is controlled by the controller resident in the machine.

This co-ordination task is sometimes referred to as sequence control (machine tools and PLCs) or flow of control (robots) and is usually concerned with the logical control of the machine system. It is also known as process control, as it has been widely applied in the process industries.

We will assume throughout that our system is largely recording its current configuration and state using discrete binary sensors.

As we move down the control hierarchy, from the higher level host computers to the machine tools and the robots in our systems, it becomes more necessary to have timely communication between the elements of the system. The control therefore becomes more real time. 'Real time' can be used to describe any information processing activity or system which has to respond to externally generated input stimuli within a finite and specified delay. An alternative informal working definition is to consider real time as being 'fast enough to be useful'.

We begin our discussion by indicating the properties we require in a computer for such control tasks, and then discuss the principles of hierarchical control. The balance of the chapter will indicate actual architectures that have been used to achieve control at various levels in the system.

10.2 Computer requirements for process control

Let us briefly review computer requirements for process control, especially for the control of real-time processes.

The computer must allow the following.

1. *Time-initiated events.* The computer must be capable of responding to events that are triggered by time – such events are the collection of management information and the recording of process variables at regular intervals.
2. *Process-initiated interrupts.* The computer must be able to respond to incoming signals from the process. Depending on their relative importance, these may require the computer to interrupt its current task to carry out one of higher priority. It must, therefore, be capable of interfacing with sensors to permit process monitoring.
3. *Computer commands to process.* For process control the computer must have the software capability to direct the hardware devices to carry out tasks. It must be able to be interfaced to actuators.
4. *System and program initiated events.* As the control computer will invariably be connected to other computers it must be able to handle events related to the computer itself, such as data transfer.
5. *Operator initiated events.* The control system software must also be capable of accepting inputs from operators and output of commands to trigger operator activities. Examples of such tasks are (a) asking the operator to input the identity of parts entering the system, or (b) to guide the operator through the correct system start-up conditions.

Control computers generally have single or multilevel interrupt systems and a real-time clock.

The clock of the computer is the internal device that co-ordinates the activities of the computer, by allowing the synchronization of events at specified internal 'clock-ticks'. The real-time clock implies that this internal clock is calibrated to be correct to time in the real world.

The priority levels of interrupts usually encountered are shown in Fig. 10.1, showing that a process interrupt is dealt with straight away while an operator interrupt is given the lowest priority. The response of single and multilevel interrupt systems is also indicated in Fig. 10.1, and this shows how the multilevel system can deal more quickly with urgent tasks.

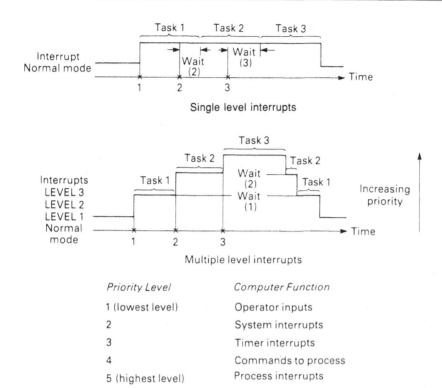

Fig. 10.1 Priority levels of interrupts and response of a computer control system to interrupts
Source: Groover (1986)

In the mechanical world it is usual for the most urgent tasks to be dealt with in a timely way to avoid irreversible damage.

The management task that we have to carry out with our control computer can be achieved by using a multivariable servomechanism or a hierarchy of computers.

10.3 A multivariable servomechanism

The inputs to a computer control system are the control commands, specifications of set points, and the state of the system as reported by the system sensors and past data which has been held in the memory. Such a control system can be represented as a feedback control loop as shown in Fig. 10.2 where the system, by operating on these inputs, generates an output to drive the system actuators.

The control activity can simply be expressed as

$$P(t) = H\ S(t - \Delta t),\tag{10.1}$$

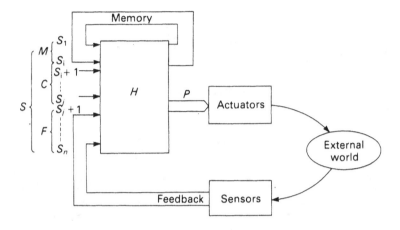

Fig. 10.2 Multivariable servo control
Source: Albus (1981)

where $S(t)$ is the row vector of the input variables, which varies with time. This vector $S(t)$ is made up of a vector C of input commands, a vector F of status feedback and M, a vector of the memory, such that

$$S = C + F + M, \tag{10.2}$$

and where $P(t)$ is the row vector of the outputs and H is the operator describing the 'rules' by which the system operates. The Δt indicates that the computational process involved takes a finite time.

10.4 Hierarchical control

It is usual to control the lower levels of a manufacturing system with a hierarchy of computers, and this control hierarchy is often practically partitioned according to organizational models similar to those outlined for the manufacturing facility in the last chapter. An example of such a hierarchy divided to allow the control of a number of cells is shown in Fig. 10.3. This hierarchical division is often forced upon manufacturing because of the range of suppliers of equipment at the lower levels and the number of machines at the lower levels. The controls of these low level machines are often tailored to obtain the best performance from their particular mechanical and electrical design. It is unlikely that this will be duplicated in higher level controllers.

In such a control hierarchy the computer at the top of the system sets the system goal. The computer at the top of the hierarchy is not necessarily the most sophisticated (recall this later when we examine the PLC). This

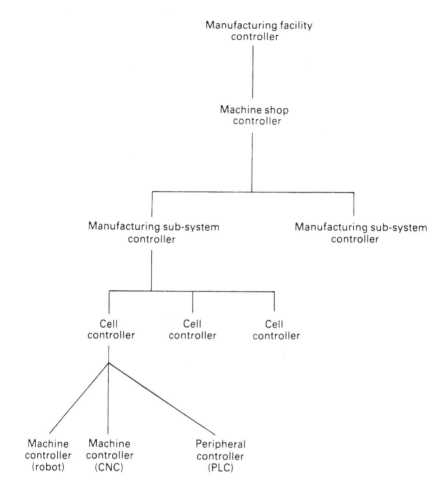

Fig. 10.3 A pragmatic control hierarchy

high level goal is devolved into successively smaller sub-goals as we go
down the hierarchy. An example of a high level system goal is a demand
to produce certain numbers and varieties of components during an operating shift – this will have to be translated into low level goals as a request
to run a particular part program at a specified time. This is sometimes
called task decomposition. At this point you should recall the discussion
of task level programming in robotics.

Albus (1981) has devised a scheme of representing the control activity
in hierarchical control, similar to that outlined above for the single level
control activity.

The architecture is built around three parallel interconnected hierarchies
as shown in Fig. 10.4. The hierarchies and modules are:

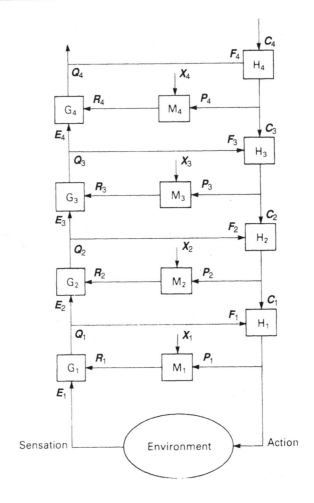

Fig. 10.4 A cross coupled processing generating hierarchy
Source: Albus (1981)

1. A behaviour-generating hierarchy which performs task decomposition on the basis of sensory information. The module associated with this is the task decomposition module, **H**, which transforms a command from the level above to a series of commands for the level below.
2. A sensory processing hierarchy which extracts the information needed by the decision-making process at each level. The sensory processing module, **G**, associated with this performs pattern recognition on sensory data.
3. A world model hierarchy which provides expectations and predictions about sensory information to the sensory processing modules. The module associated with this is a predictive memory module, **M**, which predicts what will be sensed, based on the context.

The operation of these modules can be described, in vector notation, as follows.

The task decomposition module, **H**, transforms an input vector S into an output vector P in a similar way to equation (10.1),

$$P_i(t) = \mathbf{H}_i(S_i(t - \Delta t)), \tag{10.3}$$

where the subscript i denotes the level in the hierarchy and H performs a computation with a period Δt.

The sensory decomposition module, **G**, performs a pattern recognition operation to name the sensed pattern. This may be expressed as

$$Q = \mathbf{G}\,D, \tag{10.4}$$

where D is the input vector derived from the sensory input E and a vector R from the predictive memory module. The vector R is used to condition the operation to take into account the context of the observations and sensory noise. The output vector Q is the name of the sensed pattern, which is an input to the module above the particular one in the hierarchy.

The predictive memory module, **M**, provides the sensory processing module with a prediction of what sensory data to expect. It takes as an input a vector which is composed of the P vector and an X vector which represents the context information generated by the other **G** or **H** modules. It outputs the vector R to input into D. This can be represented as

$$R = \mathbf{M}\,(P + X). \tag{10.5}$$

This sort of architecture is capable of generating complex behaviour, depending on the number of levels in the hierarchy, the number of inputs and the sophistication of the task decomposition (**H**) modules.

The model can be applied as far down the computer hierarchy as we wish. Figure 10.5 shows a theoretical task level decomposition for an assembly robot. The lower levels of the hierarchy are the most conventional, servo-control of machine axes.

In application the task decomposition is performed in less steps than is implied above. An example of this can be seen in Fig. 10.6, which shows a hierarchy for the control of a manufacturing workstation (cell) with three levels. This particular hierarchical control was implemented using a hierarchy of three state transition tables. The workstation state is read from the input sensors and is compared with the allowable states in the transition table. When a match is found the particular state transition activity to be carried out is instigated. In this way the system goes from state to state – a particular example of a finite state machine (Fig. 10.7).

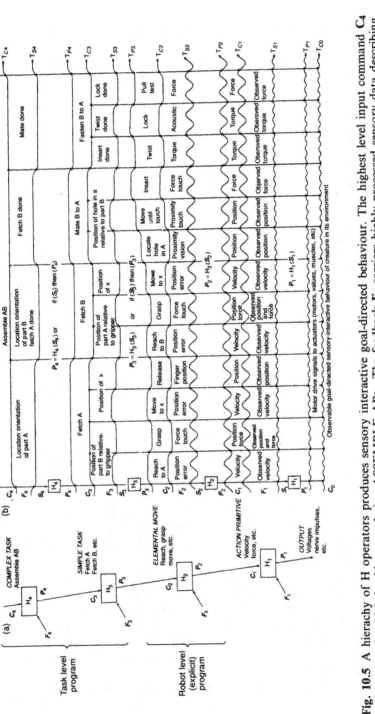

Fig. 10.5 A hierachy of H operators produces sensory interactive goal-directed behaviour. The highest level input command C_4 defines a goal, which in this example is <ASSEMBLE AB>. The feedback F_4 carriers highly processed sensory data describing the state of environment in which the assemble command must operate, including the state of the lower level P vectors. The H_4 operator maps each input S_4 into an output P_4. As F_4 changes the goal <ASSEMBLE AB> is decomposed into a sequence of subgoals <FETCH A>, <FETCH B>, <MATE B TO A>, <FASTEN B TO A>. At each level in the hierachy a different type of feedback data with a different rate-of change drives the decomposition of a higher level command into a sequence of lower level subcommands. Finally, at the lowest level the P_0 vector consists of motor drive signals which actuate observable behaviour C_0. This figure is adapted from *Brians, Behaviour and Robotics*, by J. S. Albus, © 1980 by Permission of McGraw Hill, Peterborough, New Hampshire.

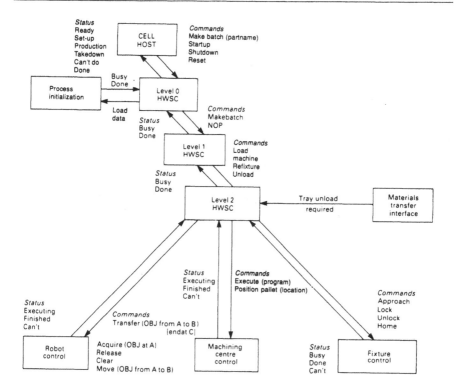

Fig. 10.6 The horizontal workstation controller
Source: Strouse *et al.*

10.5 Control networks

The control system that we have begun to to describe has a number of intelligent nodes represented at each level by a control computer. The system also generally has an overall control computer at the top of the hierarchy, often called the system host. This distribution of control and intelligence means that parts of the system will still function if other elements fail, and therefore increase system reliability. Some nodes are duplicated so that the control functions can be rapidly exchanged on machine failure (this is only usually carried out at the host computer level, and requires complex real-time data management and inserts expensive redundancy in the system).

The style of networks usually encountered in manufacturing is shown in Fig. 10.8. These are the star, the tree, the multidrop (sometimes bus) and the loop or ring. The star topology is used for very closely co-operating machines, as it is deterministic and fast. Buses and rings, however, are being increasingly used to transfer data by local area networks between remote devices that do not need to communicate in real time.

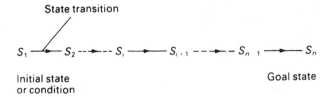

Fig. 10.7 A finite state machine

10.6 Interfacing

We will here briefly examine the techniques of interfacing computer-based devices, the interconnection of open systems and the use of local area networks

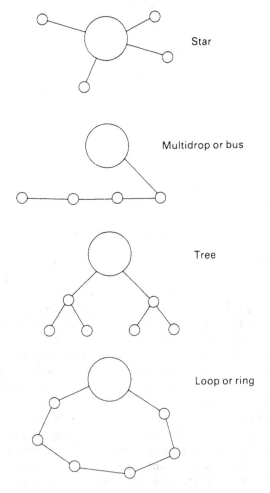

Fig. 10.8 Network topologies

within the constraints of manufacturing. You will recall that there are a number of ways of organizing a computer network (Fig. 10.8) and here we will examine the interconnections or interfaces that are usually encountered.

10.7 Data transmission

10.7.1 PARALLEL DATA TRANSMISSION

Within computers themselves information is transferred in a parallel manner (see Fig. 10.9) where a word of data is transmitted from one device to another 'byte by byte', via a 'bus', i.e. each of the eight bits of a word is transmitted simultaneously. This method is sometimes used to connect computer-based automation devices when the distance between devices is of the order of 50 cm. The connector used is a ribbon cable.

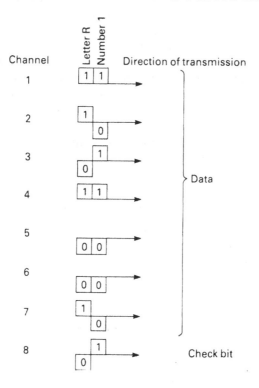

Fig. 10.9 Parallel data transmission

10.7.2 SERIAL DATA TRANSMISSION

Serial data transmission is commonly used to connect discrete computer-based devices, and data transfer is shown schematically in Fig. 10.10. The

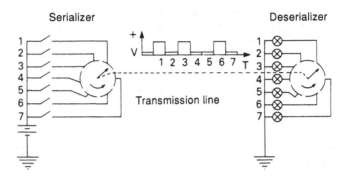

Fig. 10.10 Serial data transmission

parallel data within each computing machine must be converted into serial ('bit by bit') data, transmitted as individual bits and reconverted on receipt to the parallel data that the target computer requires.

An RS232 port is the most commonly encountered serial port and is often used to connect computers into 'star' networks. Data transmission requires 'handshaking' which ensures that the computers are able to communicate with each other before the data is transmitted. Handshaking is part of the 'communications protocol' of the devices. Protocols are standardized methods to ensure safe data transmission.

The subset of the interconnection commonly found uses only five connections. These are ground, TXD (transmitted data) the output line, RXD (received data) the input line, and RTS (ready to send) and CTS (clear to send). The TXD line of the sending device becomes the RXD line of the target device and vice versa. The protocol operates as follows:

1. the sender setting the RTS line,
2. the target senses this and sets the CTS line (the above takes between 20 and 250 ms),
3. the sender transmits the message.

10.8 Open Systems Interconnection (OSI) and LANs

An open system of computers interconnected by a data network is a system whose elements can be changed without the necessity to alter the balance of the system. This is a general principle that should be followed in the design of computer systems, however it has found greatest application in the design of communications systems, and particularly, local area networks.

There is an International Standards Organization Open Systems Interconnection (ISO-OSI) reference model that is used to aid the standardization of the layers of protocols of local area networks between computers. As we implied above, a protocol is the standardized procedure which controls data communications between devices in computer systems. A local area network (LAN) is a geographically restricted computer network that is owned by a single organization.

The seven layers of the ISO-OSI model are shown in Fig. 10.11 and the function of each layer is expanded in Table 10.1. The definitions of the layers range from at the bottom, Layer 1, the physical medium used to transmit the message such as a 'twisted pair' (of wires) or a fibre optic cable to at the top, Layer 7, the message format. These layers are embodied in proprietary networks, based on Ethernet and token rings, most of which are encountered in recent factory automation installations.

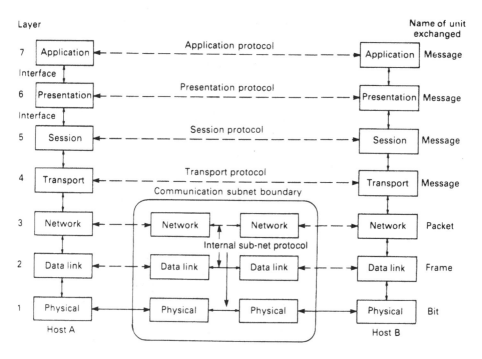

Fig. 10.11 The ISO OSI network standard

LANs are increasingly being used in manufacturing automation to pass high level control information and part programs to machines on the shop floor. They are used to interconnect equipment at a single level within the overall control hierarchy. 'Data highways' have been used for this for a number of years. LANs are not yet capable of transmitting in real time to allow the control of closely cooperating machines. It is likely that such LANs (when they do emerge) will not have the full seven layers of the OSI model.

Table 10.1 The ISO OSI layers

Layer	Layer name	Functional description
1	Physical layer	Provides mechanical, electrical, functional, and procedural characteristics to establish, maintain and release physical connections.
2	Data link layer	Provides functional and procedural means to establish, maintain and release data lines between network entities (e.g. terminals and network nodes).
3	Network layer	Provides functional and procedural means to exchange network service data units between two transport entities (i.e. devices that support transport layer protocols) over a network connection. It provides transport entities with independence from routing and switching considerations.
4	Transport layer	Provides optimization of available communication services (supplied by lower-layer implementations) by providing a transparent transfer of data between session layer entities.
5	Session layer	Provides a service of 'binding' two presentation service entities together logically and controls the dialogue between them as far as message synchronization is concerned.
6	Presentation layer	Provides a set of services that may be selected by the application layer to enable it to interpret the meaning of the data exchanged. Such services include management of entry exchange, display and control of structured data. The presentation layer services are the heart of the seven-layer proposal, enabling disparate terminal and computer equipment to intercommunicate.
7	Application layer	Provides direct support of application processes and programs of the end user and the management of the interconnection of these programs and the communication entities.

10.9 Problems with local area networks

Local area networks have problems in the environment required for the control of manufacturing facilities. Communication in manufacturing requires that all messages sent are received while they are still relevant. They are, however, being used increasingly for the passage of data such as part programs. Problems arise because there are a number of computers in contention for the network, each perhaps waiting to pass messages around the system, each needing to communicate its message in a deterministic, timely and efficient way.

An examination of Ethernet, a widely applied network, shows this problem.

10.10 Ethernet*

The Ethernet is a particular example of a carrier sense multiple access collision detect (CSMA/CD) mechanism and is shown in Fig. 10.12. At the point marked t_0 a station has finished transmitting its packet of data. A packet is a group of bits with a fixed maximum size and well defined format that is transmitted as a whole. Any other station having a packet to send may now attempt to do so.

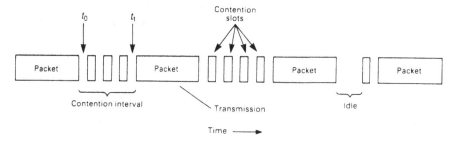

Fig. 10.12 The Ethernet mechanism
Source: Tanenbaum (1981)

If two or more stations decide to transmit simultaneously, there will be a collision. Each will detect the collision, abort its transmission, wait a random period of time, and then try again, assuming that no other station has started transmitting in the meantime.

Ethernet therefore consists of alternating contention and transmission periods, with idle periods occurring when all stations are quiet (through lack of work, for example).

Now let us look closely at the details of the contention algorithm. Suppose that two stations both begin transmitting at exactly time t_0. The length of time it takes the stations to detect the collision between the two messages determines how long the contention period will be, and therefore what the delay and network throughput will be. The minimum time to detect the collision is the time it takes the signal to propagate from one station to the other.

This collision detection is an analogue process. The station's hardware listens to the cable while it is transmitting. If what it reads back is different from what it is putting out, it knows a collision is occurring. (Hence carrier sense collision detect.)

---

* Tanenbaum (1981); adapted by permission of Prentice-Hall Inc., Englewood Cliffs, NJ.

To minimize delay, an adaptive randomization strategy has been devised. This minimizes delay under light loads and is stable under heavy loads. It works as follows. After a packet is successfully transmitted, all stations may compete for the first contention slot. If there is a collision, all colliding stations may compete for the first contention slot. If there is a collision, all colliding stations set a local parameter (L) to 2 and choose one of the next L slots for retransmission. Every time a station is involved in a collision, it doubles its value of L. In effect, after K collisions, a fraction 2^{-K} of the stations will attempt to retransmit in each of the succeeding slots. As the Ethernet becomes more and more heavily loaded, the stations automatically adapt to the load. This heuristic is called binary exponential back-off.

To examine the performance of Ethernet let the time for a signal to propagate between the two farthest stations be τ. At t_0 one station begins transmitting. At $\tau - \varepsilon$, an instant before the signal arrives at the most distant station, that station also begins transmitting. It detects the collision almost instantly and stops, but the little noise burst caused by the collision does not get back to the original station until time $2\tau - \varepsilon$. Therefore, in the worst case, a station cannot be sure that it has seized the channel until it has transmitted for 2τ without hearing a collision. This time is known as the slot. On a 1 km long co-axial cable $\tau \approx 5\,\mu s$.

Consider Ethernet under conditions of heavy and constant load, with K stations always ready to transmit. If each station transmits during a contention slot with probability p, the probability of each other station not transmitting is $(1 - p)$; the probability A, therefore, that some station acquires the ether during the slot is

$$A = K p (1 - p) \ldots (1 - p) \tag{10.6}$$

K stations attempting to transmit · · · · · probability of this single station transmitting · · · · · probability that each of the other stations is not transmitting

Rearranging this:

$$A = Kp(1 - p)^{K-1}. \tag{10.7}$$

A is maximized when

$$\frac{dA}{dp} = 0, \tag{10.8}$$

i.e.

$$\frac{dA}{dp} = [(K - 1)(1 - p)^{K-2}.\, Kp(-1)] + K(1 - p)^{K-1}$$

and

$$p = \frac{1}{K},$$

(10.9)

so that A_{max} becomes

$$A_{max} = K \cdot \frac{1}{K}\left(1 - \frac{1}{K}\right)^{K-1} = \left(1 - \frac{1}{K}\right)^{K-1},$$

(10.10)

so that as $K \to \infty$, $A \to 1/e$

$$\left(\text{recall}\left(1 - \frac{1}{n}\right)^n = \frac{1}{e} \text{ as } n \to \infty\right).$$

The probability that the contention interval has exactly j slots in it (i.e. that successful transmission occurs in the jth slot) is

$$(1 - A)(1 - A) \dots (A)$$

(10.11)

probability of not succeeding for $j - 1$ slots

probability of succeeding on the jth slot

or

$$\text{probability of } j \text{ slots} = A(1 - A)^{j-1},$$

(10.12)

so the mean number of slots per contention is given by

$$\sum_{j=1}^{\infty} jA(1 - A)^{j-1} = \frac{1}{A}.$$

(10.13)

(Recall $\bar{x} = \sum_{i=1}^{\infty} if(i)$, where i is the value of x, and $f(i)$ is the probability of x.) The mean number of slots is $1/A$ and the slot length is 2τ so that the mean contention interval is

$$W = \frac{2\tau}{A}.$$

(10.14)

Assuming optimal p, the mean number of contention slots is never more than e, so W is at most

$$2\tau e \simeq 5.4\tau.$$

(10.15)

If the mean packet takes Ps to transmit, when many stations have packets to send, the channel efficiency (the fraction of the channel used usefully) is

$$\text{channel efficiency} = \frac{\text{time for packet transmission}}{\text{time for packet transmission} + \text{contention interval}}$$

$$= \frac{P}{P + 5.4\tau} \tag{10.16}$$

Here we see where the maximum cable distance between any two stations enters into the performance figures – the longer the cable, the longer the contention interval.

As you can see, this process is probabilistic and not deterministic, as the message may not be passed within a specified time or with certainty. The network performance (Fig. 10.13) becomes less efficient with more stations and shorter messages.

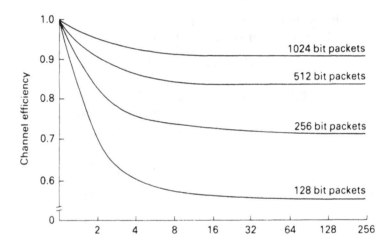

Fig. 10.13   Ethernet efficiency (10 Mb, 1 km)
Source: Tanenbaum (1981)

10.11 Token rings

One method of overcoming this lack of determinism is to use a token ring as shown in Fig. 10.14. In this kind of ring a special bit pattern, called the token, circulates around the ring whenever all stations are idle. Typically, the token will be a special 8-bit pattern, for example 11111111. Bit stuffing is used to prevent this pattern from occurring in the data. Bit stuffing, for example, inserts a 0 after 11111 in the transmitter and removes it in the receiver, this therefore prevents 11111111 occuring in anything else but the token.

When a station wants to transmit a packet, it is required to seize the token and remove it from the ring before transmitting. To remove the token, the ring interface (which connects the station to the ring (see Fig. 10.14)) must monitor all bits that pass by. As the last bit of the token passes by, the ring interface inverts it, changing the token (e.g. 11111111) into another bit pattern, known as a connector (e.g. 11111110). Immediately after the token has been so transformed, the station making the transformation is permitted to begin transmitting.

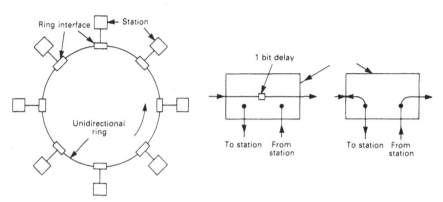

Fig. 10.14 A token ring
Source: Tanenbaum (1981)

In such a way a station will always have access to the net when its turn is due, and the message is assured of transmission. However, the time taken to acquire the network may still be too long for real-time activities.

10.12 The Manufacturing Automation Protocol (MAP)

The General Motors Manufacturing Automation Protocol is a particular example of the ISO-OSI model based on a broadband token bus, and is regarded as a secure data transmission method for the shop floor. The network using MAP was intended to be the factory 'backbone' to carry data between all the devices and cells on the shop floor. Each of the seven layers of the ISO-OSI model have been defined and Fig. 10.15 shows the standards at each level of MAP embodied in MAP Version 3. The broadband system is that used in the USA for cable TV and is a proven, rugged, multichannel communication system. It uses a co-axial cable carrying a number of frequency channels, each using a modulated carrier wave.

It is unlikely that there will be universal standard for LAN protocols throughout the local and national business and manufacturing systems of a company – MAP has evolved to include TOP for office automation. It is also felt that MAP will be unsuitable for timely communications in real time and that there will be a reduced protocol for communication between devices that need real time message passing. Therefore, there will have to

Layers	Function	Map specification
User program	Application programs (not part of the DSI model)	HMFS/EIA 1893A
Layer 7 application	Provides all services directly comprehensible to application programs	ISO case kernel
Layer 6 presentation	Transforms data to/from negotiated standardized formats	Null at this time
Layer 5 session	Synchronize and manage data	ISO session kernel
Layer 4 transport	Provides transparent reliable data transfer from end node to end node	ISO transport class 4
Layer 3 network	Performs message routing for data transfer between non-adjacent nodes	ISO CLNS
Layer 2 data link	Error detection for messages moved between adjacent nodes	IEEE 802·2 Link level control class 1
Layer 1 physical	Encodes and physically transfers messages between adjacent nodes	IEEE 802.4 token access on broadband or carrierband modulation on coaxial media
	Physical link	

Fig. 10.15 The General Motors MAP specification

be separate LANs for separate areas of the manufacturing business. These different LANs are connected by devices known as 'gateways' or 'routers' which can translate between communications protocols. LANs generally have a restricted length, and bridges join similar networks.

10.13 Databases in control

The computers in control of a manufacturing system are required to store a great deal of data (for example a monolithic FMS host has to carry 4 Megabytes). Some of this data is static, and remains unchanged for the life of the system (such as machine or robot part programs). Some data is continually changing or dynamic (such as that recording parts position in the system) and needs to be reliably accessed by a number of devices. In an ideal world this would be stored as a single database so that all computers in the system would be sure of having the same data (recall the discussion in Chapter 6). This can be achieved by having a single database as shown schematically in Fig. 10.16, or by having a logically integrated distributed database (Fig. 10.17) – this second approach has the advantage that the data within a system will reside actually where it is required. In manufacturing systems, control of such single database concepts are at the moment rarely practicable.

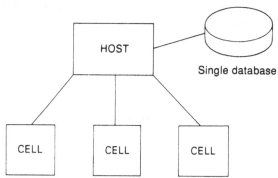

Fig. 10.16 Single system database

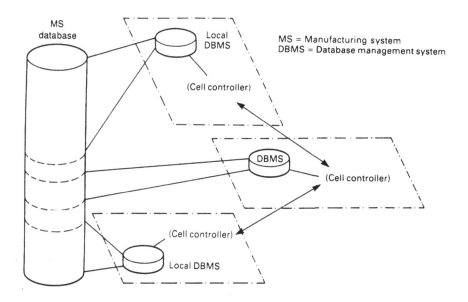

Fig. 10.17 Logically integrated distributed database
Source: Ranky (1985)

One of the most difficult problems in manufacturing systems integration is the interfacing of a wide variety of devices made by many different manufacturers. We have reviewed the physical interconnections above in our discussion on data transmission. Interfacing also requires the sharing of data and messages. One of the solutions that has been proposed to resolve part of the problem requires the use of a single database, to which each of the individual devices is interfaced. To achieve this only one software interface needs be written for each machine instead of the great number that would need to be written for direct machine to machine communication. The interfacing of all systems using a standard neutral format, e.g. the International Graphics Exchange Standard (IGES) for CAD, is a similar philosophy.

A control system architecture (AMRF) showing the concepts of hierarchical control with single static and dynamic databases is shown in Fig. 10.18.

10.14 Control hierarchy embodiment

The control hierarchies used in practice can be summarized by the following variants.

10.14.1 MONOLITHIC AND CELLULAR FMS

A typical monolithic FMS control hierarchy is indicated by Fig. 10.19. This shows the FMS host receiving schedules from another computer, and downloading part of the program data via an optical data highway (a ring network) to the individual machine controllers and system control commands to a system 'sequence' controller. The individual machine and transport system controllers feedback their status to the sequence controller, which gives them operating commands based on its interpretation of the overall system status. Note that the transport system has a separate dedicated control computer. Tooling and setting areas will also be under the control of computer systems supervised by the host. There is also a requirement to track parts and pallets around the system.

In the case of cellular FMS the individual machine controller is replaced by a cell controller.

10.14.2 FLEXIBLE TRANSFER LINE

A flexible transfer line is usually controlled using a programmable logic controller (PLC) or small computer (mini or micro) to supervise the individual machines (see Fig. 10.20). One of the essential functions of this controller is to track product variants within the line.

In robot lines the robot part program is usually resident at the machine, although more recent systems have used a host computer to store part

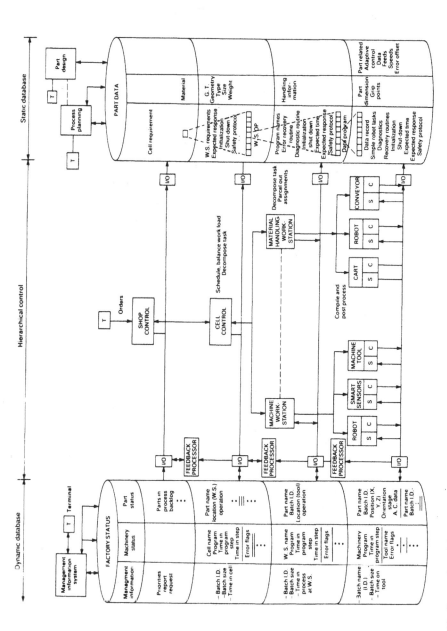

Fig. 10.18 Control architecture of the AMRF *Source:* Mclean

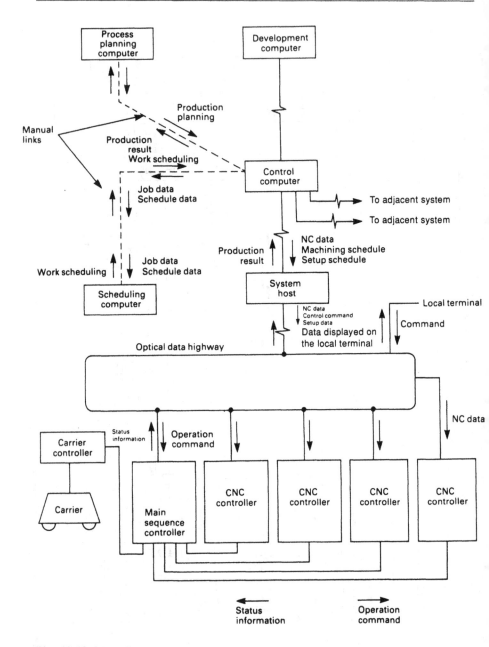

Fig. 10.19 Monolithic FMS control

programs. For CNC systems the part programs can be downloaded on, for example, a data highway to the individual machines. This is not essential to the function of the line.

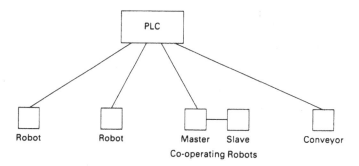

Fig. 10.20 Robot transfer line

10.14.3 FLEXIBLE CELLS

The real-time control of the closely co-operating machines in the cell can be accomplished in a number of ways:

1. using a conventional or purpose built mini- or microcomputer;
2. using a programmable logic controller;
3. using 'sequence' control capabilities resident in one of the machines within the cell.

Methods 2 and 3 are the most frequently encountered in both robot and machine tool cells. A PLC is resident in most machine tools to control the 'sequence' (M Code) activities of the machine (e.g. turn coolant on via a solenoid valve). The PLC within the controller allows 'conditional part programming', that is the selection of particular elements of the program from, for example, sensor inputs. A PLC is also often used to control robot cells. This is because the sequence control available in robot controllers is not particularly fast when dealing with a large number of inputs (see the discussion below).

For the cell control task all of these variants generally take most of their feedback from binary sensor arrangements.

10.15 Control computer embodiment

This section examines the control computer embodiment at various levels in the control hierarchy.

10.15.1 HOST AND MINICOMPUTERS

Host computers are general conventional minicomputers or small mainframe computers programmed in a conventional imperative language like

PASCAL or COBOL. They usually have a large memory to accommodate the system dynamic database and a variety of input–output facilities. More conventional computers are used because the task being carried out is more like data processing than real time control at this high level in the hierarchy. It is also easier to interface such computers to the other conventional computers used within companies for management data processing and CAD.

10.15.2 MICROCOMPUTERS

Conventional microcomputers are sometimes used for the control of system elements or cells. They are an expensive alternative to purpose built machines and are rarely fitted with sufficient input–output or operating systems suitable for real time control of a variety of devices. They are, however, more general purpose computers than PLCs and allow sophisticated calculations to be made. Microcomputers and single board computers are often used as interfaces between proprietary controllers and the balance of a system.

A number of purpose built (rugged, with a large I/O) cell control computers have begun to be marketed. These as yet are rarely installed in systems as they are regarded as unproven technology.

10.15.3 PROGRAMMABLE LOGIC CONTROLLERS

A PLC is essentially a particularly rugged, purpose built microprocessor that has been specially designed to handle a large amount of input and output, I/O, and execute 'logic' or 'sequence' control over a number of co-operating devices. They are usually programmed using a microcomputer, dumb terminal or special purpose programming unit. There are a number of high level aids to program development, some of which are discussed in the next chapter. This control program is then loaded into the PLC RAM (random access memory) or 'burnt' into EPROM (eraseable programmable read only memory) to carry out the control task.

PLCs are widely applicable to many forms of automation, and a block diagram of a large machine is shown in Fig. 10.21 demonstrating the range of input and output devices.

10.15.4 MACHINE CONTROLLERS

As has been mentioned above many machines have sequence control facilities within their controllers which can be used to control cells and their peripheral devices. Figure 10.22 shows schematically the PLC resident in a machine tool.

In most robots, however, the sequence control is carried out by the processor that supervises the co-operation of the separate machine axes

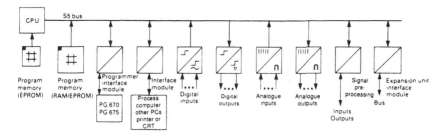

Fig. 10.21 Block diagram of PLC
Source: Siemens

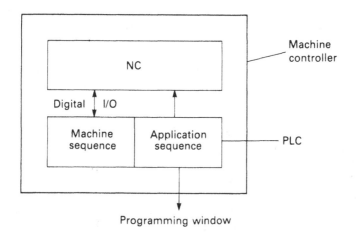

Fig. 10.22 Resident PLC in machine tool
Source: Fanuc

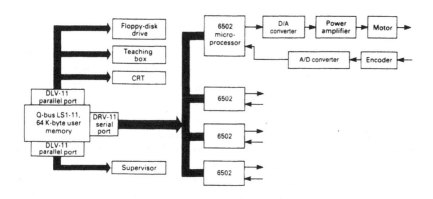

Fig. 10.23 Robot controller
Source: Unimation

(see Fig. 10.23). This concurrent processing means that the controller reaction time can be long when the machine is expected to deal rapidly with a large I/O, and in this case it would be usual to use a PLC as well as the robot controller.

Multiprocessor controllers have emerged for robots that use three separate processors within the same controller linked by a bus to a common memory (see Fig. 10.24). This is a similar solution to using a common

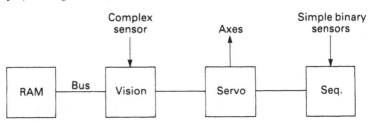

Fig. 10.24 Multiprocessor robot controller
Source: Mohri *et al.*

database. One processor is used for axis supervision, one for sequence control and one for complex sensor (vision) system control. The use of a special processor for sequence control may remove the speed problem outlined above. The use of such systems is restricted by the length of fast parallel bus that can be operated in a noisy industrial environment.

It has been suggested that at this level in the hierarchy the processor controlling the high level sensing information will be the supervisor of the other devices, even though they are connected as a parallel processor array. Vision is also the area where large purpose built parallel processor arrays (one processor per pixel, for example) to execute particular image analysis algorithms quickly are likely to be first applied commercially in manufacturing.

10.15.5 HETERARCHIC ARCHITECTURES

The architecture used for the control of manufacturing is currently a fashionable research area; it has been suggested that the operation of such systems will become increasingly parallel and that heterarchical architectures will evolve. Heterarchic, 'all on the same level', distributed control architectures imply that all machine controllers will show increasingly intelligent behaviour, and be able to communicate rapidly in real time. The more intelligent controllers will, for example, bid for work instead of being allotted work by a high level host. Pragmatically, such architectures will not be applied in manufacturing for some time. Flat hierarchies, with few levels, will be both practical to install and fast and reliable to operate for the foreseeable future.

11 Manufacturing systems software

After reading this chapter the reader should understand:

- software tasks in manufacturing system control;
- real-time features of manufacturing control;
- typical languages for small system control;
- emerging software methods.

11.1 Introduction

This chapter will examine the software aspects of manufacturing systems and cells. It begins by briefly examining the sorts of tasks that software is required to carry out in a manufacturing system and a manufacturing cell. As many of the activities in controlling such cells and systems are real time, the chapter then continues by examining the design requirements for real time languages. The chapter will then close by examining in some detail programming languages for the computers at the lower levels of the control hierarchy, such as those that are used for cell and small system control.

The treatment of high level manufacturing software given here is superficial because of our concentration on the low level hardware – the machines that carry out the processes that add the value. The reader is directed to the further reading section for a more complete treatment of production planning and control (PP&C/CAPM) issues – PP&C applications tend to tackle the complexity inherent in manufacturing systems that have either grown organically or make products with a large variety of parts, such as airframes.

11.2 Software tasks in manufacturing systems

Most manufacturing software is prepared in modules, each module being linked to others. If it is necessary to tailor any of the modules to the particular system it is only the relevant modules that are modified rather than a very large single software system. This technique allows these modules to be used in a number of systems to defray the extremely costly software development process. It is said that professional applications programmers can only prepare 10 new lines of code per working day. Such code has to be reliable and well documented, and all the effects of new code on old code must be taken into account. Table 11.1 shows the software size of some manufacturing systems. This does demonstrate the software overhead associated with programmable systems.

The sections below indicate the contents of typical software modules.

Table 11.1 Software complexity in robot systems
Source: Carlisle

Example	*Size*
University of Michigan generic part sorting cell	30 000 lines 4 CPUs
Adept demonstration at Robots 8 exhibition	6000 lines 2 robots 2 vision systems 1 conveyor
Martin Marietta intelligent task automation project	75 000 lines 8–10 CPUs 5–6 languages
Automated monitor line	700 000 lines 2 hosts
Schaffner electronic assembly system	16 000 lines in robot controller and host PLC 10 other PLCs

11.3 Scheduling

Software is needed to define the schedule of work (the 'what shall we make today?' decision) to be carried out within the production system. This can be downloaded to the system by a 'management' computer but is more usually generated interactively by a human system controller. A simple off-line simulation of the system capacity is usually used for this.

If an AGV-based transport system is used the activities of the vehicles have to be scheduled. This is carried out using simple scheduling rules known as dispatching rules. Rules that have been used for a two machine cell with an AGV are as follows.

1. Load the machine with longest cycle time.
2. Empty the load station and allow another component to be loaded in the system.
3. Feed the other machine.

The system also needs some ability to change these scheduling rules in the event of a machine breakdown, for example.

Such re-scheduling can be regarded as an error recovery activity. Such activities are required at all levels in the system hierarchy. At the lowest level, the error detection and recovery routines must be able to prevent permanent damage to the machines.

Error recovery routines are generally explicitly programmed contingency actions. Robot assembly systems often have many such routines for example, to cope with parts absence in feeders and excessive insertion forces.

11.4 Generation of management information

Software must exist to allow the collection of data representing system performance which can be used for management control, such as system output, machine downtime and tool lives.

11.5 Part program communication

There must also be facilities to allow the communication of part program data around the system and interfacing to the part program generating system, whether it is a manual or a remote CAD system. This is especially important in machining and sheet metal working systems or cells. DNC technologies have been applied in these situations for a number of years.

It has not been usual to have such facilities in robot systems, but DNC-type systems are emerging for the storage and transmission of robot part programs. These have been installed primarily to secure storage of robot part programs. Robot part programs are particularly sensitive to alterations, and alterations are comparatively easy to carry out. Central program storage with limited access prevents unauthorized program modification.

11.6 Data and databases

Database software to handle the storage of data within the system must be rugged and secure. The sorts of dynamic and static data that must be stored in a flexible manufacturing system, for example, are:

1. a tool library which contains the number and type of individual tools in the system and their position in the system;
2. a record of individual tool offsets, the geometrical deviations of each tool from the ideal tool used within the general NC part program;

3. a record of fixture data, the position and identity of fixtures within the system, the parts that can be fixtured using them and any modifications that need to be made to tailor a general fixture to a particular part;
4. part programs;
5. current part position data recording software that keeps track of the individual identity and positions of parts within the system;
6. current pallet data recording facilities that keep track of full and empty pallets and moving and stationary pallets.

The pallet and part tracking activity is made more complex because each pallet may hold more than one type of component at once, and usually also holds more than one of each component. Pallet identification is usually accomplished using binary sensor arrays to read pin arrays on the pallet, or by reading transponders resident on the pallet.

11.7 Interfaces to other sub-systems

The system must have software to allow it to interface easily and understandably with operators who may have to load and unload components and set tools for the system. It must also communicate with automatic tool control and transport systems if these are present.

11.8 Control of closely co-operating machines

Software is needed to manage the interactions of closely co-operating machines and monitor simple sensors and the machine status. This requires the ability to carry out conditional tasks, either in the system controller program or in the machine part-program. When, for example, two robots are working within the same workspace it will be necessary for the two machines to be aware of 'no-go' areas that change with time.

11.9 Machine monitoring

The condition of the individual machine has to be monitored, e.g. coolant and lubrication flows, temperatures, and cutting forces must be monitored and the machine shut down if set point conditions are not met. Diagnostic expert systems are now being applied to support this task.

11.10 High level sensor interfaces

At the lowest system level software is required to change a machine path with respect to inputs from high level sensors. Examples of this are the use of touch trigger probes in machining systems to modify NC part programs with respect to real part orientation, and vision systems for robot systems which select or modify particular part programs.

11.11 Design requirements of real-time languages

Many of the elements above can be real time control activities. It is therefore useful to review the design requirements of such languages.

11.11.1 SECURITY

The security of a language design is a measure of the extent to which programming errors can be detected by a compiler or language run time support system. Exhaustive software testing is generally not possible for large software systems.

A compiler prepares a machine language program from a program written in high or problem level language, which is subsequently executed – whereas an interpreter does this for each instruction which is then executed immediately.

Real time application development in robotics is much easier and faster with an interpreted language when compared to a compiled language. The compiled program, however, is likely to be faster on final execution.

11.11.2 READABILITY

The language must be readable, i.e. easily understood by only examining the program text. This, as well as requiring a readable language using common constructs (if. then. else., etc.), needs the programmer to write in a structured way.

Typically a structured program is a hierarchy of modules – each having a single entry point and a single exit point; control is passed downward through the structure without pathological connections (goto . . .) to higher levels of the structure.

Readability has a number of benefits: it is easier to write reliably in readable and structured languages; it reduces documentation as the program is the central piece of documentation; and it is therefore easier to maintain.

11.11.3 FLEXIBILITY

The language must be flexible enough to allow the programmer to carry out whatever he wishes, without having to use machine code inserts or other similar techniques. Machine language inserts are fast to execute, can be very versatile but are however very difficult to write. Language flexibility is hard to achieve in real-time control, as there is often the necessity to control a wide variety of non-standard peripheral devices.

11.11.4 SIMPLICITY

The language must be simple to learn and use. Simplicity not only arises from the commands themselves but is ensured by having minimum restriction on the use of these commands. It is easy to remember the individual commands, but difficult to remember the conditions and restrictions on them.

11.11.5 PORTABILITY

The language design must be independent of the underlying hardware, and this should allow the language to be moved from machine to machine. This is difficult to achieve in real time control, where the essential task of the language is to extract the maximum benefit from the underlying hardware.

A real-time portable language that is becoming increasingly encountered in manufacturing is C running under the UNIX^{TM} operating system; it is, however, much more usual to find that each device and manufacturer has a different language. Recall the range of robot languages and the varieties of subsets of APT.

11.11.6 EFFICIENCY

Real time systems must often achieve high computational throughput in order to meet the constraints imposed by the system being monitored or controlled. The constraints are generally those of timely and deterministic operation. The language used must therefore be fast and efficient and have consistent response times.

The language must also be easy to write, so that programs can be generated efficiently, and these two different efficiency requirements are sometimes contradictory.

The discussion below shows some of the languages that have been used in the real-time control of manufacturing cells and systems, though many of them do not compare favourably with the ideal requirements above.

11.12 Programming languages for system control

Computers high in the hierarchy usually use the more familiar programming languages such as COBOL, PASCAL or FORTRAN. This is because the tasks that they carry are more like conventional data-processing activities than the real-time control tasks at the lower levels. High level control computers are also required to be interfaced eventually to other factory systems, such as CAD or MRP (materials requirements planning) systems. These are usually written in similar languages.

11.13 Cell control computer languages

Table 11.2 shows some of the instruction set of a recently introduced system level computer.

Table 11.2 A cell computer instruction set
Source: Gould

Category	Instruction name
Booleans	AND, OR, XOR, INVERT, R/S and J/K FLIP FLOPS (logical 'set' operations)
Timers	1/100, 1/10 and 1 second intervals (timing with specified interval)
Counters	UPCOUNTER, DOWNCOUNTER (incrementing and decrementing variables)
Mathematics	ADD, SUBT, MULT, DIV, MOD, SQRT, SIN, COS, TAN COT, LOG, LN, 10**X, e**X, Y**X, Xth RT Y (arithmetic, trigonometric and exponential operations)
Assignment	SET, RESET, ASSIGN, COPY (variable definition operations e.g. SET variable EQual number)
Logical	GT, GE, EQ, LE, LT, NE (variable tests, e.g. variable Greater Than (GT) count)
Bit/Matrix	SENSE, SET, CLR, AND, OR, XOR, INVERT, COMPARE (two matrices) (measure, and manipulate bit and matrix data, e.g. bits representing sensor data)
String	SUBSTR, SRCHSSTR, DELSUBSTR, STRLEN (string (alpha-numeric message) manipulations, SUBstitute, SeaRcH, DELete, measure LENgth)
Serial I/O	ENCODE (construct message), TERMIO (transmit/receive serial data), DEVIO (read/write list of variables from a programmable device)
Other	IF, FOR, SELECTOR

This included Boolean logic operations, three timers, upward and downward counting facilities, mathematical manipulations, the ability to assign variables, conventional logic operations (greater than, etc), matrix and single bit manipulations, string (STR) manipulation and message manipulations (ENCODE, etc.) and IF, FOR constructs.

As you can see, this is very like a conventional computer language but does include Boolean operations for the logical control activity. The matrix operations allow, for example, the manipulation of machine tool offsets and robot transformation matrices. The string and message manipulation commands are particularly important when controlling lower level devices. It is often required to manipulate and generate strings of code in the language of the lower level device to allow the whole system to perform in an optimal way.

11.14 Programmable logic controllers

The most frequently encountered cell and line level control computer is the programmable logic controller (PLC).

There are essentially three methods of programming a PLC in its machine language, using the graphic aids of a ladder logic diagram and control flow diagram. Programs are prepared on the machine programming unit and downloaded into the PLC. Machine language programs are directly transmitted to the machine but programs prepared by the graphics aids have to be compiled into the machine language.

The machine language is made up of statements, and these are very similar to the machine code encountered for microprocessors and micro-computers and machine tool code instructions. Such statements are executed very quickly. The program or statement list is executed sequentially. The statement has two elements:

- Operation: 'what is to be done?'
- Operand: 'what is it to be done with?'

The operand also has two elements:

- Operand identifier (input, output, etc.)
- Parameter (e.g. the particular number of the I/O).

The signal state at the input is usually binary, that is either '1' or '0' ('high' or 'low'). In operation the PLC scans the signal states at its inputs and executes logic operations in accordance with the statement list.

Typical operands and some typical operations are shown in Tables 11.3 and 11.4. These lead to a statement list in the form shown below (source: Siemens) statements can be stored as program blocks and manipulated conditionally.

Table 11.3 PLC operands
Source: Siemens

Operand	Meaning	Operand	Meaning
I	Input	FB	Flag byte
Q	Output	PB	Peripheral byte
F	Flag	KT	Constant times
T	Timer	KC	Constant counter number
C	Counter	KF	Constant fixed point number

(A flag is where the processor stores a signal state.)

Table 11.4 PLC operators
Source: Siemens

Operator	Meaning
A	Scan for '1' and AND with result of previous logic operation
AN	Scan for '0' and AND with result of previous logic operation
O	Scan for '1' and OR with result of previous logic operation
S	Set
R	Reset
=	Assign '1' if result of logic operation '1' (similarly '0' and '0')
CU	Increment counter
CD	Decrement counter
SP	Start timer
L	Load value into accumulator
! = F	Comparison
JU	Jump unconditionally
JC	Jump conditionally

```
      .
   A   I 2.0
   JC  =4
3: O   I 1.0
   =   Q 0.2
   R   0 0.3
   .
   JU  =5
4: A   I 1.1
   =   Q 0.3
   R   Q 0.2
   .
   R   F 4.3
5: A   I 2.1
   =   Q 1.1
```

If there is a '0' signal at input 2.0 continue to program block 3, if there is a '1' at input 2.0 jump to program block 4: program block 3 says; provided that there is a '1' signal at input 1.0 set output number 0.2 to '1' and reset output number 0.3: program block 4 says; if input number 1.1 and the previous logic operation are '1' set output 0.3 to '1', reset output 0.2 and reset flag 4.3: program block 5 says; if input 2.1 and the previous logic operation are '1' set output 1.1 to '1'.

The other main method of preparing a PLC program is to use the graphic aid of a ladder logic diagram, as shown in Fig. 11.1. Such diagrams are usually used by less expert programmers. Ladder logic is a traditional method of constructing relay based logic. The rungs of the ladder connect the 'power' supplied along the right hand rail to the 'ground' of the left hand rail, and in this way the outputs and relay 'coils' are energized. A coil is a method of storing an intermediate value in the program.

11.15 Flow of control commands in robots

Robot high level languages include a number of 'flow of control' commands for logical control. Many manufacturing cells are built around robots and it is sensible to use the robot controller as a cell controller.

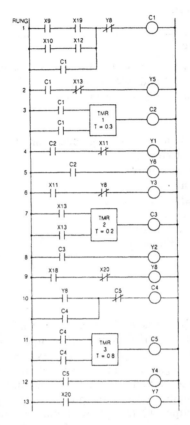

Execute:

Rung 1
((If input X9 and input X19 on) or (If input X10 and input X12 on) or (coil C1) and not output Y2 set then activate coil C1 to store (input X9 and X19 on or . . . and not output Y2)))

Rung 2
(If coil C1 activated and input X13 not on then set output Y5 on)

Rung 3
(If coil C1 on and the timer has been reset by coil C1, start timer 1, after 0.3 turn on coil C2)

.

Rung 13
(. . . .)

go back to rung 1

Fig. 11.1 A ladder logic diagram
Source: Texas Instruments

Although such languages are easy to program in, they do have some speed limitations.

The early approach to the flow of control can be shown using VAL as an example. The listing below (from Unimation) shows some of the flow of control commands in VAL, the Unimation language. These allow the interrogation of sensors and conditional actions depending on the sensor state. VAL is an interpreted language and allows fast application development.

```
        SETI COUNT = 0
10      SETI COUNT = COUNT + 1
        IF COUNT GT 5 THEN 20
        GOTO 10
20      IFSIG 2, -3,,THEN 30
        .
        .
        .
        SIGNAL 1
        .
        REACT 3, ROBOT
        .
        WAIT -1
        .
        .
        .
```

Set the integer count to zero, increment it by one, if it is greater than five go to label 20, if not go back to label 10, at label 20 test input signals 2 and 3: if 2 is positive and 3 negative go to label 30, . . ., set output 1 high, . . ., react to signal 3 positive by executing the subroutine ROBOT, . . ., don't do anything (WAIT) until you get a negative signal on input 1.

Perhaps the most characteristic member of the new generation of more structured robot languages is VAL II (from Unimation) and enhanced VAL II (from ADEPT). These languages allow input from high level sensors to alter the path of the robot in real-time and also support a separate PLC program. An example of this (from Unimation) is shown below.

```
.PROGRAM CIRCLE
 1 PROMPT 'Input radius in mm',rad
 2 TYPE 'Move to centre of circle'
 3 TYPE 'Press COMP button when done'
 4 DETACH ;Allow user to use teach pendant
 5 ;Wait for COMP button on teach pendant
 6 DO
 7 UNTIL PENDANT(2) BAND 20
 8 ATTACH ;regain control of arm
 9 DECOMPOSE c[] = HERE ;get xyz data for centre
10 TYPE/B, 'Moving in 10 seconds' ;beep terminal
11 DELAY 10
```

```
12 PCEXECUTE pc.alt, -1,0 ;start ALTERing program
13 ;set internal ALTER, WORLD mode cumulative
14 ALTER(-1,19)
15 WHILE SIG(1032) DO ;Continuously make circles
                                ;until signal stops it
17     FOR ang = 0 TO 360 STEP 5
18         x = rad*COS(ang) + c[1]
19         y = rad*SIN(ang) + c[2]
20         MOVES TRANS(x,y,c[3],c[4],c[5],c[6])
21     END
22 END
23 PCEND ;Finish up altering program
 .END
```

```
.PROGRAM PC.ALT
1; This program will ALTER the z component of the
2; circle, dependent upon inputs from the external
3; binary signal input lines 16-21
4;
5; E.G. If the binary bits are set up like:
6; BITS 1021 1020 1019 1018 1017 1016
7;       0    1    0    1    1    0
8;
9; Then this corresponds to a 22
10; correction in the z direction. Because
11; this value will be divided by two (see ALTOUT)
12; the correction would be 11 mm.
13;
14 ALTOUT 0,0,0 BITS (1016,6)/2,0,0,0
```

(There are two programs here; the program CIRCLE which draws circles, the first of which is defined by the operator, until an input tells it to stop; and a program PC ALT, called by the first program, which monitors external signals and modifies the z position of the circle, these programs run concurrently.)

You will see that the programs are structured and commented, try and translate them for yourself, the ALTER command allows modification to robot co-ordinates in real time in response to, for example, high level sensors.

The GMF robot language KAREL, from GMF (General Motors Fanuc) allows sophisticated flow of control and structured programming using commands such as those indicated in Fig. 11.2.

Another well-known robot and cell language AML II (A Manufacturing Language II), from IBM, has very similar flow of control commands as shown in the demonstration program in Fig. 11.3. AML is a compiled language.

```
ALTERNATION CONTROL STRUCTURES

    1., IF expression THEN
          {statement}
        ENDIF

    2.  IF expression THEN
          {statement}
        ELSE
          {statement}
        ENDIF

    3.  SELECT expression OF
          CASE ( constant <, constant> : {statement>
          {CASE ( constant <, constant> : {statement}}
          <ELSE : {statement}>
        ENDSELECT

LOOPING CONTROL STRUCTURES

    1.  FOR Id = expression TO expression DO
          {statement}
        ENDFOR

    2.  FOR Id = expression DOWNTO expression DO
          {statement}
        ENDFOR

    3.  REPEAT
          {statement}
        UNTIL expression

    4.  WHILE expression DO
          {statement}
        ENDWHILE

UNCONDITIONAL BRANCHING

    1.  GOTO Id

    2.  GO TO Id

CONDITION HANDLER STATEMENTS

    1.  CONDITION [ expression ] :   -- Condition handler definition
          WHEN conditions DO actions
          {WHEN conditions DO actions}
        ENDCONDITION

    2.  ENABLE CONDITION [ expression ] -- Turn on the monitoring

    3.  DISABLE CONDITION [ expression ] -- Turn off the monitoring

    4   PURGE CONDITION [ expression ] -- Delete a condition  handler
        definition

    5.  SIGNAL EVENT [ expression ] -- Signal an event condition
```

(A condition handler is a set of conditions to be monitored in parallel with program execution, actions may interrupt normal program execution sequences)

Fig. 11.2 Flow of control commands in KAREL which allow structured programming

```
 1 /* THIS IS A TEST APPLICATION PROGRAM SHOWING A NUMBER OF
 2     THE FEATURES OF AML2.   THESE INCLUDE LINEAR AND CIRCULAR
 3     MOTION,USE OF PALLETS AND USE OF THE SENSOR COMMAND.
 4     THIS PROGRAMME IS WRITTEN FOR A 7576 MANF CONTROL SYSTEM.
 5 */
 6 pt1:new pt(300,-300,-100,180);    ##Define points used in the program
 7 pt2:new pt(300,300,-50,-180);
 8 pt3: new pt( -40., -300.);
 9 pt4: new pt( 5, -340 );
10 pt5: new pt(  300, 300,0,3600);
11 pt6: new pt( -400, 300,-100,0);
12
13 pts:new pt(-200,600,0,0);
14 pts1:new pt(-40,440,0,0);
15 pts2:new pt(120,600,0,0);
16
17 ll:new pt(-700,380,0,0);          ##Define the corners of the pallet
18 lr:new pt(-700,180,0,0);
19 ur:new pt(-460,180,0,0);
20
21 s:new sensor(17,5);               ##Define a multi-point sensor
22
23  speed=accel=decel=1.;            ##set motion variables
24
25 test2:subr()
26 pal:new pallet(ll,lr,ur,4,20);
27          i:new 0;
28          release();
29          while ++i le 9 do begin
30          pmove(pt(300,300,0,0));
31          get_a_part();
32          setpart(pal,i);
33          getpart(pal);
34          release_a_part();
35          end;
36 end;
37
38
39 test3: subr();
40          i:new 0;
41          while i le 3 do begin
42              i=i+1;
43              ELBOW=LEFT;
44              pmove( pt3);
45              ELBOW=RIGHT;
46              pmove( pt4);
47          end;
48      ELBOW=EITHER;
49      pmove( pt5);
50      pmove( pt6);
51 end;
52
53 get_a_part:subr()
54          zmove(-50);
55          grasp();
56          delay(0.1);
57          zmove(0);
58 end;
59
60 release_a_part:subr()
61          zmove(-50);
62          release();
63          delay(0.1);
64          zmove(0);
65 end;
66
67
68 /* The main subroutine starts after this comment
69    note that the name of the subroutine is the same
70    as the programme name.  Programme execution starts
71    from this point.*/
```

```
72
73  test76:subr()
74  circular(0);
75  writeo(s,1);            /* Set sensor to 1 */
76  pmove(pt1);
77  zmove(-100);            /* Lower Z shaft */
78  grasp();               /* Close gripper */
79  zmove(0);              /* Raise Z shaft */
80  writeo(s,2);            /* Set sensor to 2 */
81  pmove(pt2);
82  writeo(s,3);            /* Set sensor to 3 */
83  movetype=straight;     /* Do a straight move */
84  pmove(pt1);
85  movetype=natural;      ## Lets do a circular move
86  pmove(pts);            /* Move to start of circle */
87  circular(1000);
88  pmove(<pts1,pts2,pts>); /* Do circular move */
89  circular(0);
90  movetype=straight;
91  elbow=either;          /* Reset elbow */
92  movetype=natural;
93  test3();
94  test2();               /* Lets do a pallet */
95  end;
```

Fig. 11.3 An AMLII program for the IBM 7535 Manufacturing System
Source: IBM

From the variety of this small sub-set of the robot languages you will see the need for some form of standardization and recall IRData, between the high level language and the manipulator.

11.16 Rule driven systems

Rule driven systems can be applied in the essentially logic-based activity of machine management, and such systems are based on sets of IF condition THEN action rules, and are sometimes called production systems. These are extensible and easy to write – a typical rule (a condition – action rule, from Bourne (1986)) is shown below. You will notice that KAREL allows the use of this programming technique. Such activities are comparable with the task level programming indicated in the machine programming chapter.

condition	action
(antecedent)	[consequent]
if	then
(And (Hold billet)	[(Move robot door)
(Open door)	(Move robot furnace)
(Vacant space))	(Move billet)
	(Move robot door)
	(Close door)]

(IF the robot is holding a billet and the furnace has an open door and there is a vacant space in the furnace, THEN move the robot to the furnace door and move the robot into the furnace and place the billet in the vacant space and move the robot back to the door and close the furnace door.)

One particular style of rule-based system based on fuzzy sets (in fuzzy set theory membership of the set need not be 1 or 0) has many applications in manufacturing control. Fuzzy control uses rules based on the operator's experience of the behaviour of the system and allows decisions to be made on inexact information.

11.17 The future

As more and more processes yield to programmable approaches, and as systems become larger, more complex and increasingly sensory interactive, manufacturing control and programming are particularly important and fast-moving areas. The years since the preparation of the first edition of this book have reaffirmed that people are an essential part of the manufacturing system. Over the next few years we are likely to see many developments in manufacturing systems, ranging from drives for increased standardization of the integrating technologies to the applications of techniques developed by the computer scientists in artificial intelligence. One of the particularly important techniques to be applied will be object orientation in support of both the software design process and code reuse. Object oriented software systems use closed software elements containing both data and procedures that communicate solely by messages. Efforts in all these areas, and the technologies between these extremes, will, without doubt, be reflected in the competitiveness, prosperity and long-term survival of those companies and industrialized countries prepared to solve their manufacturing problems using integrated, computer-assisted and controlled systems.

12 Examples

These examples are included to support the academic reader. One of the purposes of the book is to act as a 'primer' on manufacturing systems, particularly for those whose specialization during the whole of an undergraduate course is not manufacturing engineering.

The examples are divided into two sections. The first section is intended to support the student as their understanding of the technologies and design issues evolves. The second section is made up from modified questions set by the author when examining on the final year of the Production Engineering Tripos (now Manufacturing Engineering Tripos) at Cambridge University. These questions are intended to build upon the candidate's understanding of engineering science and to test the candidate on how well he or she can take an overview of the whole topic. The final test of a manufacturing engineer is the effectiveness of the choice of manufacturing system configuration used to make a generation of products and the ability of this choice to accommodate the next generation of product — these questions push the candidate to focus on these issues, the design of the manufacturing system.

In many of the questions the candidate is asked to compare technologies. Perhaps the most effective way of answering such questions is to describe briefly the key parameters that are to be used in the comparison and then to present a table identifying how well each technology satisfies these parameters. Close the answer to the question by summarizing the trade-offs.

12.1 Manufacturing systems

The questions in sections 12.1–7 are set to support the development of understanding. They may build on engineering science knowledge that is not in the book.

12.1.1 Prepare an IDEF0 model of the way this book attempts to help the reader to improve their understanding of manufacturing engineering. Draw a 'single box' level 0 model (like Fig. 1.6), expand this into three sub-models (as Fig. 1.7) and expand one of these sub-models into a further three sub-models.

12.1.2 Define and discuss:

(a) WIP
(b) door-to-door time
(c) just-in-time
(d) batch-of-one
(e) right-first-time
(f) ship-to-stock
(g) synergy in manufacturing systems
(h) functional integration

(Answers: Careful review of the text will find most of the answers, ship-to-stock is the ability to place purchased components into inventory or on to the production line *without inspection*, this implies that suppliers have assured product quality.)

12.1.3 Examine equation (2.3) describing a Kanban system.

(a) Identify the true variables for a system made up of well understood discrete processes.
(b) Establish what the system with $\alpha = 0$, $C = 1$ and T_w represents.
(c) How can these values be approached in conventional facilities?

The Kanban philosophy was developed in Japanese automotive facilities during a time when their market was rapidly expanding; how appropriate is it for other types of manufacturing facilities?

12.1.4 Given the GT coding systems for turned components in Fig. 2.8, due to Optiz, what is the GT code for the component in Fig. 12.1? Indicate alternative processes that could be used to manufacture the product and indicate the machines that are likely to be included in a GT cell to make the part. (Answer 13 232.)

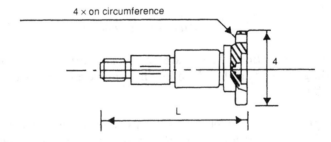

Fig. 12.1 Turned component

12.2 Machine tools, robots and vision

12.2.1 Define and discuss:

(a) loss of form;
(b) robot workspace;
(c) serial and parallel robot arms;
(d) unmanned and minimally manned;
(e) offline programming.

12.2.2 In an experiment with a machine tool the frequency response curve was obtained for excitation of a particular mode of vibration. The peak disturbing force was 200 N and the amplitude of vibration at resonance was 0.1414 mm. Resonance occurred at 50 Hz and the frequencies giving an amplitude of 0.1 mm were 45 and 55 Hz. Estimate the viscous damping constant, ζ, from the relationship, when $x = x/2$ at ω_1 and ω_2, for small values of ζ

$$(\omega_1 - \omega_2) = 2\,\omega_n\,\zeta.$$

Also estimate the equivalent spring stiffness, k, for the system from the frequency responses shown in Fig. 12.2. Recall that the static deflection of the system is amplified by the resonance.

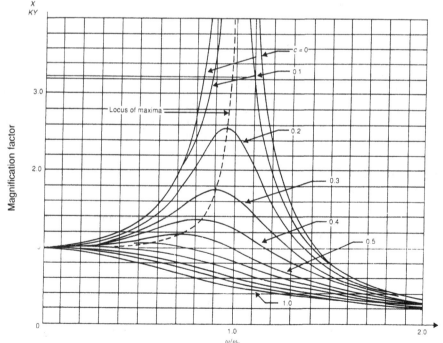

Fig. 12.2 Frequency response curves

Using equation (3.14) calculate the critical chip width for a material with cutting coefficient of 150 kg/mm^2 when the cut overlap factor, μ, is one.

(Answer $\zeta = 0.1$, $k = 6.8 \times 10^6$ N/m, $b_{cr} = 10$ mm.)

12.2.3 What are the prerequisites for scenes that are easy to analyse with machine vision?

Indicate the steps required to extract the position and orientation of the object shown in Fig. 12.3.

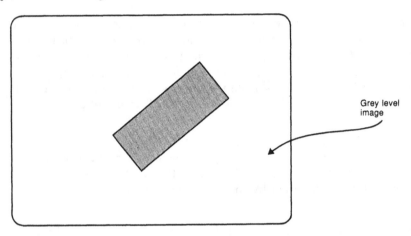

Grey level image

Fig. 12.3 Simple grey level image

12.2.4 Robots with vision are now frequently used in the glazing of car windscreens as is shown in the schematics contained in Fig. 5.20. This installation uses a separate computer controlled gripper held by a large manipulator unable to accept path modifications from a sensor. The gripper contains a number of servo-systems to move the screen relative to the aperture and an integral vision system. Figure 12.4 shows a similar system constructed from a robot able to accept path modifications from a sensor. What are the major differences between the systems?

12.2.5 Sketch the skeletons of the robots shown in Fig. 12.5 and 12.6. You may scale from the diagrams if necessary.

What are the major differences between the robot manipulators?

12.3 Automated cells and systems

12.3.1 Examine the suitability of industrial robots and CNC machining centres for inclusion in minimally manned programmable manufacturing systems. Con-

sider both hardware and software aspects and present your answer as a table
of key points.

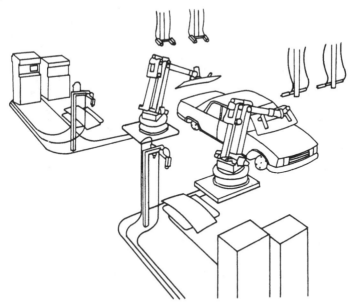

Fig. 12.4 An alternative arrangement for robot glazing

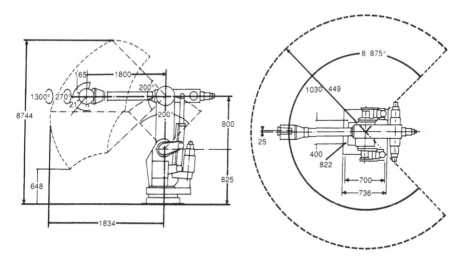

Fig. 12.5 Anthropomorphic robot schematic

Compare the values of the CNC machine tool and robot when applied
in isolation. Focus on the added value from each machine and the manner
in which each machine reduces product lead time.

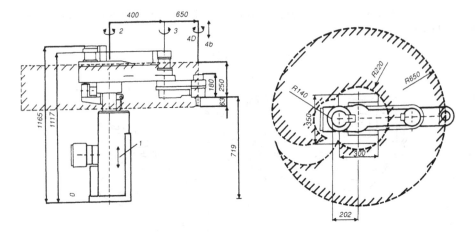

Fig. 12.6 SCARA robot schematic

12.3.2 Briefly define flexibility.

Compare the systems shown in Fig. 9.6 and 9.7 with respect to their flexibility and 'fitness-for-purpose' (appropriateness to the task). Think carefully about the products that the systems make and therefore the processes that they contain.

12.3.3 It is proposed to use a cylindrical co-ordinate robot to handle parts through a machine tool cell. The cell is to have an output of 100 000 parts per year and will at least consist of:

- one rough drilling machine, cycle time 50 s;
- one fine drilling machine, cycle time, 35 s;
- on pull broaching machine, cycle time, 25 s;
- one input magazine.

Each part has to go through each process. Discuss alternative options for the method of operation of the cell examining particularly the range of options for scheduling parts through the cell. Comment on the cycle times achievable in the cell (noting that the speed of a robot is about the same as a man) and on whether the cell should be double or single shifted, identifying the assumptions you have made. What effect does increased demand have on the operation of your cell?

Indicate the cell control options that are available to the cell designer and select one, noting the reasons.

12.3.4 Compare the design of programmable machining systems for turned parts and for prismatic parts.

12.3.5 What are the advantages and disadvantages of cellular manufacturing systems when compared to monolithic systems?

12.4 Assembly and assembly systems

Answering the questions in section 12.4 will be helped by having access to a car workshop manual that includes a number of exploded views.

12.4.1 Find an exploded view of a carburettor and identify how many design for assembly rules it breaks. How can the product be redesigned to ease its assembly without changing its functional design?

12.4.2 A car manufacturer has to design a large assembly system to assemble engine blocks (pistons, conrods and cranshafts, etc.) and cylinder heads. The cylinder heads must then be assembled to the blocks; the other devices must also be assembled to the outside of the engine (water pumps, alternators, pulleys and carburettors, etc.).

Part of the engine block assembly and part of the head-to-block assembly can be carried out using purpose-built semi-automatic transfer machines. Such machines can process all six variants of the basic engine.

The manufacturer wishes to produce 1500 engines a day. There are six basic engines with 90 variants. Each engine takes one hour to assemble, 10 minutes of which is carried out by the semi-automatic transfer machines.

Indicate the significant problems in the design and outline a possible solution based on the use of AGVs after taking a close look at Fig. 4.2. Indicate any assumptions you have made and remember that the system you design must transport the completed engines and the parts that create those engines.

12.4.3 An assembly operation is to be made between a steel peg and a steel bore of nominal diameter 15 mm with a 0.1 to 0.2 mm clearance between the two. There is also a 1 mm wide (*W* on Fig. 8.8) 45° chamfer at the corner of the bore.

The operation is to be carried out using a SCARA robot which at the particular point in its workspace has a stiffness in its *x-y* plane of 225N/mm and about its *z* axis of 16 500N m/radian. The distance of the centre of rotation of the peg from the peg tip is 20 cm. The coefficient of friction for steel on steel is 0.2 and the robot repeatability is ± 0.05mm. Draw the jamming diagram for the system and determine the machine wedging force and moment on the peg.

(Answer: 225 N and 1089N m.)

12.5 Control and programming of machines and systems

12.5.1 Compare the real time languages used in manufacturing with the design requirements for such languages.

12.5.2 (a) It is required to interface *n* devices. This can be accomplished in two ways — by interfacing each device to each other or by interfacing each device to a neutral intermediary as is shown in Fig. 12.7.

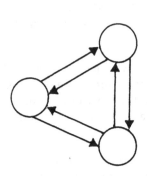
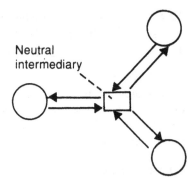

Neutral
intermediary

Fig. 12.7 Interface arrangements

By considering the amount of interfaces, *n*, that must be prepared (the neutral strategy implies that each device must have a receiving interface and a transmitting interface) determine when each strategy is suitable. Comment on the addition of a new device to the network.

(b) Sensors can be unreliable. How many parallel sensors are required sensing the same activity to detect sensor failure and to provide an absolutely reliable result?

(Answer: (a) $n < 3$ direct, $n > 3$ neutral intermediary, (b) 2, 3.)

12.5.3 Attempt to identify the optimal architecture (a hierarchy or the number of levels in a hierarchy) in a manufacturing control system by considering the maximum time taken to transmit a message in a system with *S* stations at its bottom level. The message is to be passed from one end to the other of the bottom level.

Each message passing task takes *t*, and each level of the hierarchy (there are *n* levels in the whole hierarchy) has a constant fan out *b*. Each of these definitions is repeated in Fig. 12.8 to help you.

(Answer: This problem remains a research problem. Attempting to structure the answer formally will support the student in understanding the problem — if anyone approaches an answer that will withstand robust criticism, they should consider publishing it!)

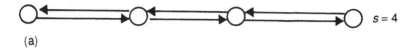

(a)

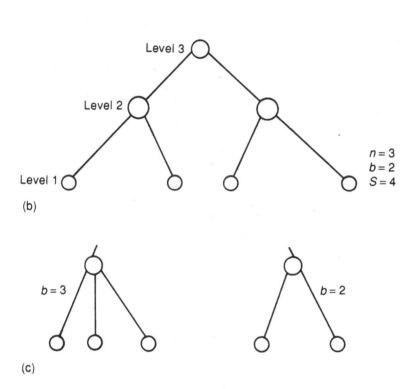

$n = 3$
$b = 2$
$S = 4$

(b)

(c)

Fig. 12.8 Definitions for hierarchical control: (a) heterarchy, (b) hierarchy and (c) fanout

12.5.4 Figure 12.9 shows the control task of a robot interfaced with a vision system using a microcomputer.

The task that the robot is carrying out is to place two markers in a scene to calibrate the robot with respect to the vision system: then parts are placed, by an operator, in the scene, to be identified by the vision system; these are then removed by the robot. Blob isolate and blob size are the mechanisms used to identify each component.

Satisfy yourself that you understand the control task represented by the net: + shows a conditional action. Using coins as tokens and moving them around to places as conditions are satisfied will help!

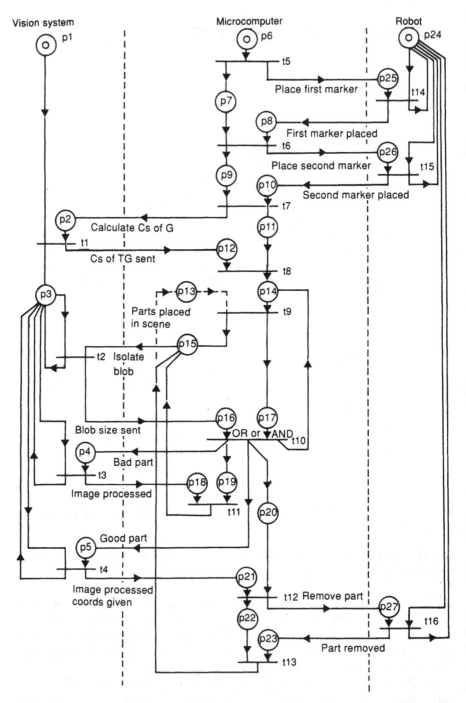

Fig. 12.9 Petri net showing a control task

12.6 Simulation

12.6.1 Describe the principles of operation of a three-phase discrete-event simulator. Compare two- and three-phase simulation (two-phase simulation is distinguished from three-phase simulation as it does not include execution of 'conditional' actions, all actions are solely based upon the passage of a specified time interval).

12.7 CIM

12.7.1 By this time the reader should have a view of 'integrated manufacturing'. The reader is encouraged to draw a linked model (diagram!) of the computer-based technologies used in CIM. The diagram should show:

(a) the discrete technologies (acronyms) applied in CIM with some comment on their significance,
(b) an understanding of the speed of operation of the individual parts of the system, identifying which parts of the system are required to operate in 'fast' real time and which parts do not have such time constraints. (What is real time?);
(c) the use of a static database (once per design), a dynamic database (once per individual product created) and a common database for parts description;
(d) the areas of the model in which there is significant human/manual input — with an explanation of why this input remains; and
(e) an indication of the major feedback loops to control the business.

Because of the approach of the question, your diagram will be 'bottom up' and be based upon a 'functional integration' view of the technological solutions to business problems. How does such a model fit with the 'top-down' business-led view of such problems?

12.8 Examination questions

These are typical examination questions and force the candidate to take an overview.

12.8.1 (a) Why has group technology been so influential in the design of successful present day factories, both manual and automated? Identify the advantages and opportunities it provides and the constraints it imposes.

(b) What is usually understood by CADCAM and CAPP? Indicate the changes that feature-based design systems are likely to impose on these.

12.8.2 By considering a production plant as a conveyor system develop an expression that relates the number of Kanban, y, in a system to the demand on that system per unit time, D, the waiting and processing time of the Kanban, T_w and T_p respectively, and the container capacity C. You should also include a policy variable α, the amount of extra inventory allowed.

Identify the significance of each variable, giving an indication of their magnitudes. Discuss how the Kanban system can be used to measure the performance of the manufacturing plant and drive it to operate as a just-in-time system. What other advantages does the system have? Does it have any disadvantages?

12.8.3 One framework for the management and instigation of change in manufacturing can be summarized by 'simplify–automate–integrate'.

Discuss each of the individual techniques, giving an example of their application. Explain why the techniques are ordered as they are.

Such phrases and many acronyms are viewed as fashionable ways of tackling manufacturing problems. Discuss why this is so, why it is unsatisfactory and present your own view of a better methodology.

12.8.4 (a) Describe the function of a direct numerical control (DNC) system and possible communications media for such a system. Prepare an outline software design for such a DNC system, noting the key characteristics of each software module.

(b) Indicate briefly why the link between computer aided design and numerical control is so important in many manufacturing processes.

(c) Discuss how hierarchical control and local area networks allow the integration of individual computer-based devices into systems.

12.8.5 Identify the tasks that the control software of a flexible transfer line as shown in Fig. 9.6 must carry out.

Draw, as a set of IDEF/SADT diagrams, the decomposition of the control software into modules. Identify the function of each of the software modules and the data that is likely to be passed between these modules.

In your decomposition take account of the computer that each module is likely to be run on and the language in which it is likely to be written.

12.8.6 Discuss the problems associated with the construction of programmable manufacturing applications, including both sensing and actuation, by outlining the design of an automatic (robot) 'ping-pong' player. Concentrate on the generic problems and solutions that such a well-understood application allows you to explore.

In your discussion highlight how any constraints you have placed on the system resolve design issues.

12.8.7 Indicate the key features that identify favourable candidate tasks for the application of industrial robots.

An industrial robot is a flexible reprogrammable manipulator. Discuss why so little of this flexibility can be exploited, giving examples from typical robot applications.

In what way do you think that the application of machine vision will increase the flexibility of future robot applications? What other major technological advances must be made to exploit the advantages of the robot manipulator and machine vision when combined?

12.8.8 (a) Why is group technology a significant enabling technology for current automated factories?

(b) Which elements of the present generation of programmable manufacturing systems are flexible and which require extensive dedicated engineering?

(c) Discuss where and why redundancy occurs in programmable machines and programmable systems for manufacturing.

12.8.9 (a) How does hierarchical control allow manufacturing devices to be integrated; further, how does it allow task decomposition for manufacturing?

(b) Describe a practicable hierarchy of computers for controlling an automated small batch manufacturing facility. What activities are carried out at each level? What kind of instruction set is necessary to enable these activities to be carried out?

(c) Computer numercially controlled machines and industrial robots are frequently the devices at the bottom of such a hierarchy. Briefly compare the methods for programming these devices.

12.8.10 Discuss the factors that must be taken into account in a considered decision to install or not install a complex automation project.

Further reading

The items included in this bibliography have three forms: books that expand on the material outlined in the book, review papers which report the state of the art, and papers which include original work that it is difficult to find in text books. This section has been considerably revised in this new edition.

Chapter 1
Introduction — manufacturing systems approaches

STRATEGY CONTEXT

Clark K. and Fujimoto T. (1991) *Product Development Performance*, Harvard Business School Press.
Porter M. E. (1990) *Competitive Advantage of Nations*, Free Press.
Skinner W. (1985) Manufacturing: missing link in corporate strategy. *Manufacturing, The Formidable Competitive Weapon*, Wiley.

MANUFACTURING SYSTEMS, ETC.

Bignall V. *et al.* (ed.) (1985) *Manufacturing Systems: Context, Applications and Techniques*, Basil Blackwell/Open University.
Chrysallouris G. (1992) *Manufacturing Systems, Theory and Practice*, Springer Verlag.
Hitomi K. (1979) *Manufacturing Systems Engineering*, Taylor and Francis.
Lucas Engineering and Systems (1991) *Mini-Guides: The Lucas Manufacturing Systems Engineering Handbook*, Lucas Engineering and Systems.
Lupton T. (ed.) (1986) The management of change to advanced manufacturing systems, in *Human Systems*, IFS Publications and Springer Verlag.
Parnaby J. (1979) Concept of a manufacturing system. *International Journal of Production Research*, **17**, 123–35.
Rembold U. and Dillman R. (eds) (1986) *Computer Aided Design and Manufacturing*, Springer Verlag.
Wu B. (1992) *Manufacturing Systems Design and Analysis*, Chapman & Hall.

CONCURRENT ENGINEERING

Nevins J. L., Whitney D. E. and colleagues (1989) *Concurrent Design of Products and Processes, A Strategy for the Next Generation of Manufacturing*, McGraw Hill.
Parsaei H. P. and Sullivan W. G. (eds) (1993) *Concurrent Engineering*, Chapman & Hall.

QUALITY

Brumbaugh P. S. and Heikes R. G. (1992) Statistical quality control. *Handbook of Industrial Engineering*, Wiley Interscience.
Spenley P. (1992) *World Class Performance Through Total Quality*, Chapman & Hall.

Chapter 2
More conventional approaches to factory layout

BASIC OPERATIONAL RESEARCH TOOLS

Wild R. (1980) *Production and Operations Management*, Holt Reinhart and Winston.

GROUP TECHNOLOGY

Burbidge J. L. (1979) *Group Technology in the Engineering Industry*, Mechanical Engineering Publications.

JUST-IN-TIME AND OTHER PRODUCTION CONTROL METHODS

Bauer A., Bowden R., Browne J., Duggan J. and Lyons G. (1994) *Shop Floor Control Systems: From Design to Implementation*, Chapman & Hall.
Monden Y. (1993) *Toyota Production System. An Integrated Approach to Just-in-Time*, 2nd edn, Chapman & Hall.
Sugimori Y. *et al.* (1977) Toyota production system and Kanban system – Materialisation of just-in-time and respect-for-human system. *International Conference of Production Research, Tokyo*, p. 553.
Vollman T. E., Berry W. L. and Whybark D. C. (1984) *Manufacturing Planning and Control Systems*, Irwin.

Chapter 3
The machining centre – a servo-controlled machine tool

MACHINE TOOLS

Boothroyd G. (1981) *Fundamentals of Metal Machining and Machine Tools*, McGraw Hill.
Weck M. (1984) *Handbook of Machine Tools*, Wiley.

MACHINE TOOL CONTROL

Koren Y. (1983) *Computer Control of Manufacturing Systems*, McGraw Hill.

Chapter 4
The robot — a handling device, a manipulator

ROBOT DESIGN AND APPLICATION

Engleberger J. (1980) *Robotics in Practice*, Kogan Page.
Fu K. S., Gonzalez R. C. and Lee C. S. G. (1987) *Robotics, Control, Sensing, Vision and Intelligence*, McGraw Hill.
Groover M. P. (1986) *Industrial Robotics: Technology Programming and Applications*, McGraw Hill.
McCloy D. and Harris M. (1986) *Robotics: An Introduction*, Open University Press.
Warneke H. J. and Shraft R. D. (1982) *Industrial Robots, Applications Experience*, IFS Publications Ltd.

AUTOMATED GUIDED VEHICLES

Muller T. (ed.) (1983) *Automated Guided Vehicles*, IFS Publications Ltd.

KINEMATICS

Featherstone R. (1983) Position and velocity transformations between robot end-effector co-ordinates and joint angles. *International Journal of Robotics Research*, **2**(2), 35–45.

GRIPPERS

Chan F. Y. (1982) Force analysis and design considerations of grippers. *Industrial Robot*, December, 243–249.

Chapter 5
Sensing

GENERAL SENSORS

Tlusty J. and Andrews G. C. (1983) A critical view of sensors for unmanned machining. *Annals of CIRP*, **32**(2).

FUNDAMENTALS OF MACHINE VISION

Barrow H. G. and Tenenbaum J. M. (1981) Computational vision. *Proceedings of the IEEE*, **69**, 572–95.
Davies E. R. (1984) A glance at image analysis. *CME*, **31** (12), 32–5.
Fu K. S., Gonzalez R. C. and Lee C. S. G. (1987) *Robotics, Control, Sensing, Vision and Intelligence*, McGraw Hill.

VISION AND ROBOTS IN THE FUTURE

Lee M. H. (1989) *Intelligent Robotics*, Open University

COLOUR VISION

Gershon R. (1985) Aspects of perception and computation in color vision. *Computer Vision, Graphics and Image Processing*, **32**, 244–77

Chapter 6
Software for single machines

NC PROGRAMMING

Koren Y. (1983) *Computer Control of Manufacturing Systems*, McGraw Hill.

PROCESS PLANNING

Chang T. C. and Wysk R. (1985) *An Introduction to Automated Process Planning Systems*, Prentice Hall

ROBOT PROGRAMMING

Lozano-Perez T. (1983) Robot programming. *Proceedings of the IEEE*, **71**, 821–40.

CAD AND GEOMETRIC MODELLING

Woodwark J. (1986) *Computing shape*, Butterworths.

COMPUTER INTEGRATED MANUFACTURE

Ranky P. G. (1985) *CIM*, IFS Publications Ltd.
Weatherall A. (ed.) (1992) *Computer Integrated Manufacture*, 2nd edn, Butterworth Heinemann.

RAPID PROTOTYPING

Kruth J. P. (1990) Material Incress manufacturing by rapid prototyping techniques. *CIRP Annals*, **39**, 2.

FEATURE BASED DESIGN AND PROCESS PLANNING

Case K. and Gindy N. (1993) Features, special issue of the *International Journal of Computer Integrated Manufacturing*, **6**, 1–162.

Chapter 7
The manufacturing cell — the building block of systems

FINANCIAL JUSTIFICATION

Thuesen G. J. (1992) Project selection and analysis, in Section IV.B Engineering economy of the *Handbook of Industrial Engineering*, Wiley Interscience.

ALTERNATIVE CONFIGURATIONS

Williams D. J. and Rogers P. (1991) *Manufacturing Cells: Control Programming and Integration*, Butterworth Heinemann.

GEOMETRIC SIMULATION

Craig J. L. (1989) *Introduction to Robotics*, Addison Wesley.

PETRI NETS

Peterson J. L. (1981) *Petri Net Theory and the Modelling of Systems*, Prentice Hall.

Chapter 8
Assembly

MECHANICS

Whitney D. E. (1982) Quasi-static assembly of compliantly supported rigid parts. *Journal of Dynamic Systems, Measurement and Control*, **104**, 65–77.

DESIGN FOR ASSEMBLY AND ASSEMBLY AUTOMATION

Boothroyd G. *et al.* (1982) *Automatic Assembly*, Marcel Dekker.
Redford A. and Lo E. (1986) *Robots in Assembly*, Open University Press.

ELECTRONICS ASSEMBLY

Edwards P. (1991) *Manufacturing Technology in the Electronics Industry*, Chapman & Hall.

Chapter 9
The integrated factory and systems

DNC

Hancock D. H. J. (1982) The design of a DNC system for use in the production of small prismatic parts. *23rd Machine Tool Design and Research Conference*, UMIST.

FLEXIBLE MANUFACTURING SYSTEMS

Greenwood N. R. (1988) *Implementing Flexible Manufacturing Systems*, Macmillan Education.
Ranky P. G. (1983) *Design and Operation of FMS*, IFS Publications Ltd.
Willamson D. T. N. (1967) System 24 — A new concept for manufacture. *Proceedings of the 8th Machine Tool Design and Research Conference*, Pergamon Press.

PEOPLE AND FACTORY AUTOMATION

American Machinist Special Report 787 (1986) People and automation. *American Machinist and Automated Manufacturing*, June.
Lupton T. (ed.) (1986) *Human Factors*, IFS Publications and Springer Verlag.

Morrison D. L. (ed.) (1992) Human factors in CIM. *International Journal of Computer Integrated Manufacturing*, Special Issue, **5**, 53–142.

Shimotashiro S. *et al.* (1984) LAN based distributed model of production control systems. *IFIP Working Conference on Modelling Production Management Systems*, Copenhagen.

DISCRETE EVENT SIMULATION

Pidd M. (1984) *Discrete Event Simulation in Management Science*, John Wiley and Sons.

Pritsker A. A. B. (1986) *Introduction to Simulation and Slam II*, 3rd edn, Wiley.

Talavage J. J. and Hannam R. G. (1988) *Flexible Manufacturing Systems in Practice: Applications, Design and Simulation*, Marcel Dekker.

Zeigler B. P. (1976) *Theory of Modelling and Simulation*, John Wiley and Sons.

THE FACTORY OF THE FUTURE

Ruff K. (1985) Contempory manufacturing systems integration. *Proceedings of NFS Workshop on Manufacturing Systems Integration*, St Clair, Michigan.

Sata T. (1984) A view of the highly automated factory of the future. *Robotics and Computer Integrated Manufacturing*, **1**, 152–9.

Warneke H. J. (1993) *The Fractal Company, A Revolution in Corporate Culture*, Springer Verlag.

Womack J. P., Jones D. T. and Roos D. (1990) *The Machine That Changed The World*, Rawson Associates.

Chapter 10
Computer control of manufacturing systems

HIERARCHICAL CONTROL

Albus J. (1981) *Brains, Behaviour and Robotics*, Byte Books.

Albus J., Barbera A. and Nagel R. (1981) Theory and practice of hierarchical control. *23rd IEEE Computer Society International Conference*, September.

NETWORKING

Rembold U. *et al.* (1977) *Computers in Manufacturing*, Marcel Dekker.

Tanenbaum A. S. (1981) *Computer Networks*, Prentice Hall.

SYSTEM CONTROL AND INTEGRATION

Jones A. and McClean C. (1986) A proposed hierarchical control model for automated manufacturing systems. *Journal of Manufacturing Systems*, **5**, 15–26.
Williams D. J. and Rogers P. (1991) *Manufacturing Cells: Control Programming and Integration*, Butterworth Heinemann.

Chapter 11
Manufacturing systems software

PLANNING AND SCHEDULING SYSTEMS

Bauer A., Bowden R., Browne J., Duggan J. and Lyons G. (1994) *Shop Floor Control Systems: From Design to Implementation*, Chapman & Hall.
Wu B. (1992) *Manufacturing Systems Design and Analysis*, Chapman & Hall.

REAL TIME LANGUAGE FUNDAMENTALS

Young S. (1982) *Real-time Languages, Design and Development*, Ellis Horwood.

ROBOTIC AND CELL LANGUAGES

Blume C. and Jacob W. (1983) *Programming Languages for Industrial Robots*, Springer Verlag.
Bourne D. A. (1986) CML: A meta-interpreter for manufacturing. *AI Magazine*, **7** (4), 86–96.
Lozano-Perez T. (1983) Robot Programming. *Proceedings of the IEEE*, **71**, 821–40.

SYSTEM PROGRAMMING FUTURES

Cox B. J. (1986) *Object Oriented Programming: An Evolutionary Approach*, Addison Wesley.
Efstathiou J. (1989) *Expert Systems in Process Control*, Longman.
Kaewert J. W. and Frost J. M. (1990) *Developing Expert Systems for Manufacturing, A Case Study Approach*, McGraw Hill.
Wright P. and Bourne D. A. (1988) *Manufacturing Intelligence*, Addison Wesley.

Index

Page numbers in **bold** refer to figures.